Genetic models of
sexual selection

Genetic models of sexual selection

PETER O'DONALD, Sc.D.

LECTURER IN GENETICS, UNIVERSITY OF CAMBRIDGE

CAMBRIDGE UNIVERSITY PRESS

CAMBRIDGE
LONDON NEW YORK NEW ROCHELLE
MELBOURNE SYDNEY

Published by the Press Syndicate of the University of Cambridge
The Pitt Building, Trumpington Street, Cambridge CB2 1RP
32 East 57th Street, New York, NY 10022, USA
296 Beaconsfield Parade, Middle Park, Melbourne 3206, Australia

First published 1980

Photoset and printed in Malta
by Interprint (Malta) Ltd

Library of Congress Cataloguing in Publication Data
O'Donald, Peter.
Genetic models of sexual selection.

Includes bibliographical references and index.
1. Sexual selection in animals.   2. Biological models.   I. Title.
QL761.036   591.5′6   78-73249
ISBN 0 521 22533 7

Contents

Preface

Darwin had explained the evolution of elaborate sexual displays and nuptial plumage by his theory of sexual selection. He thought the males would compete for the females and the females exercise a choice between the males. Male competition, or 'law of battle' as Darwin called it, obviously occurs in many species. It easily explains the evolution of male aggressiveness, the male's greater size and strength and his weapons for fighting. In polygynous species only the strongest and most dominant males may succeed in mating. This aspect of Darwin's theory has been widely accepted. But his theory of female preference in favour of the most highly adorned males was denied until very recently: it was generally thought to imply the existence of an highly developed aesthetic sense by which the females discriminated between the males. Julian Huxley (1938a, b) in two influential articles tried to show that 'display may often be of advantage to the species ... any resultant selection will therefore come under the head of Natural Selection, not Sexual Selection in Darwin's sense'. This implication of group selection is explicit in some of Huxley's examples. Although Huxley had thus confused natural selection with the implausible process of group selection, even Lack (1968) quoted Huxley's views as valid criticisms of Darwin's theory.

Well before the publication of Huxley's articles, however, Fisher in *The Genetical Theory of Natural Selection* (1930) had explained the evolution of mating preferences and thus resolved one of Darwin's main difficulties. Darwin had not been able to explain the origin and persistence of female preferences for particular male characteristics. Challenged on this point, he was, for once, reduced to mere assertion and could produce no evidence to support his views. In Darwin's theory, mating preference was a premise, but in Fisher's theory an outcome, of the evolutionary process.

The 'Genetical Theory' was the first general analysis of the effects of natural selection on genes in Mendelian populations. But Fisher attempted no such analysis in his chapter on sex and sexual selection: he set up no genetical models of sexual selection; he described his theory of the evolution of mating preferences in terms Darwin himself might have used. To describe genetic models of sexual selection and to analyse their evolutionary consequences is my purpose in writing this book. Some of the models have been put forward in previous publications. Some are original to this book.

I have first attempted, in my introductory chapter, to outline the main developments in the history of ideas about sexual selection, starting with Darwin's *The Descent of Man, and Selection in Relation to Sex* (1871). In reading Darwin, I have been astonished at the contrast between the cogency and clarity of Darwin's views and the confusion in those of his critics. Darwin had also clearly anticipated ideas that are now claimed to be original. Fisher's theory of the evolution of female preferences was the only significant addition to the theory and a necessary link in the chain of deductive inference. Fisher also claimed that the mating preferences would then select for further developments in sexual characters, leading to a 'runaway process', and giving rise to more and more exaggerated displays and nuptial plumage until eventually adverse natural selection would bring the process to a halt. The idea of a runaway process was originally Darwin's but it gains no support from Fisher's theory. Fisher had shown that the evolution of mating preferences would continually increase the intensity of selection for particular characters, but he had not shown that more extreme developments of those characters would also be selected. To explain how the females select for the more extreme males, another theoretical link is required. I postulate that the females will respond to the more extreme males as to a 'supernormal stimulus' and prefer them to the others. This extra link would then complete the chain of inference that constitutes the theory of sexual selection. But without a genetic model connecting cause and effect, the theoretical links would all be insecure. How can we be sure that a gene for a mating preference, which only has selective advantage through the male offspring of preferential matings, will be selected in subsequent generations, after genetic segregation and recombination has passed the gene to males who lack the advantageous character as well as to those who possess it? As we shall see, mating preferences will often evolve as Fisher suggested, but do not always do so; and they evolve at rates very different from the geometric rate which Fisher had deduced on the basis of his non-genetical model. It should be obvious that a genetic model is an essential starting point of any evolutionary theory.

The models of sexual selection are based on observations of mating behaviour and mating choice described in Chapter 2. This chapter is largely the result of dicussion with my late friend, Dr R. K. Murton, who died just before the first draft of the book was complete. In Chapter 3, I derive the frequencies of the matings that would take place according to the different possible mechanisms of mate selection. The genetic models are analysed in the remaining chapters. Chapter 4 is based entirely on previously published analyses of models in which a certain proportion of females exercises a complete preference in favour of particular males. Chapter 5, on models of partial expression of female preference, describes original results in sections 5.1.1 and 5.2.1. Chapter 6 on frequency-dependent preferences is original. Chapter 7 summarizes published work on sexual selection in monogamous birds; section 7.6 is original. Chapter 8 is again original, describing results obtained by computer modelling of Fisher's theory of the evolution of mating preferences. In Chapter 9, I give the previously published results of an analysis of a model of sexual selection combined with assortative mating (Karlin and O'Donald, 1978). Section 9.3, in which I fit the model to data, is original to this book, however. I finally reach the conclusion in Chapter 10 that genetic models, based on Darwin's and Fisher's theories, will satisfactorily explain the exaggeration of differences between the sexes in their reproductive rôles and the many observations of variation in mating choice. The frequency-dependence so often observed in numbers of matings achieved by different males is an inevitable consequence of the females preferring to mate with one type of male or another.

Each model of preferential mating is analysed genetically as follows: first, sexual selection for dominant and recessive characters; then, sexual selection for particular genotypes; and finally, natural and sexual selection for either dominants or recessives. In each case, I find the recurrence equation relating the gene frequency in one generation to that in the next, and hence the equilibrium frequency and the stability of the equilibrium. The general method of analysis is exemplified in Chapter 3 using a simple model of male competition. The mathematics is elementary and should easily be followed by anyone prepared to work through simple algebra. Mathematicians may find scope for a more exact treatment of the stability conditions and the rates of approach to equilibrium; they may also find that analytical expressions can be derived for equilibria in models with multiple alleles.

In Chapters 4, 5, 6 and 9, I describe how data may be fitted to the models and analysed by χ^2. My own computer programs can be made available to those who have data they may wish to test for goodness of fit. Thus, although the book is largely a record of the research I have been engaged in for the past three years, it is also intended to be a handbook of methods that the experimental population geneticist or ethologist can use if he wishes to design experiments and fit his data to particular models of mating behaviour. It is in no sense, however, a review of the literature in the field of sexual selection and population genetics. I have made no attempt to find and review the literature of sexual selection in different groups and species of animals. But I have read carefully some of the earlier contributions which I regard as important in the history of ideas about the theory and tried to judge how significant they were. So often have the early contributors, including Darwin and Fisher, been misunderstood and misrepresented, that some authors of review articles can scarcely have even glanced at the works they were reviewing. Such errors I have tried to avoid – by omission!

It is with a deep sense of loss that I acknowledge my debt to the late Dr R. K. Murton. Many of my ideas about sexual selection and mating behaviour had their origin in our long and enjoyable discussions. While visiting Stanford University, I discussed some of the genetic models with Professors S. Karlin and M. W. Feldman; their precise, clear views on population genetics greatly helped me to clarify my own. At the suggestion of Professor Karlin I analysed the models of sections 5.1.1 and 5.2.1; Chapter 9 is the direct result of our collaboration. To Professor Karlin I offer my thanks for the financial support for my visit to Stanford. And I am most grateful to Dr A. W. F. Edwards for carefully reading and commenting upon the whole typescript of this book.

1

Introduction: development of the theory of sexual selection

1.1 Darwin's theory of sexual selection

In *The Origin of Species*, Darwin (1859) gave a concise definition of sexual selection and its mechanism:

This depends, not on a struggle for existence, but on a struggle between males for possession of the females; the result is not death to the unsuccessful competitor, but few or no offspring.

Darwin thought that the males' chances of mating would depend not only on competition with their rivals, but also on female choice: sexual selection would thus promote the evolution of strength, courage and weapons for fighting and also adornments for sexual display to females.

This theory, which Darwin only briefly discussed in *The Origin of Species*, is analysed in detail in his later book *The Descent of Man, and Selection in Relation to Sex* (1871).

In *The Descent of Man*, Darwin used the theory to explain many examples of the evolution of sexual dimorphism in animals. He recognized, of course, that many sexual differences would have evolved by natural selection and was clear in the distinction he drew between natural selection and sexual selection:

When the two sexes differ in structure in relation to different habits of life, they have no doubt been modified through natural selection and by inheritance limited to one and the same sex. So again the primary sexual organs, and those for nourishing and protecting the young, come under the same influence.

Sexual selection may operate according to Darwin:

When the two sexes follow exactly the same habits of life, and the male has sensory or locomotive organs more highly developed than those of the female, it may be that the perfection of these is indispensible to the male for finding the female; but in the vast majority of cases, they serve only to give one male an advantage over another, for with sufficient time, the less well endowed males would succeed in pairing with the females; and judging from the structure of the females, they would be in all other respects equally well adapted for their ordinary habits of life. Since in such cases the males have acquired their present structure, not from being better fitted to survive in the struggle for existence, but from having gained an advantage over other males, and from having transmitted this advantage to their male offspring alone, sexual selection must here have come into action. It was the importance of this distinction which led me to designate this form of selection as Sexual Selection.

Thus, although prehensile organs may have evolved by natural selection for the male to hold the female, if they also help to 'prevent the escape of the females before the arrival of other males, or when assaulted by them, these organs will have been perfected by sexual selection'. The prehensile organs would thus have evolved both by natural selection and by sexual selection.

It would be difficult to add to the cogency and precision of Darwin's distinction between Natural and Sexual Selection. A character necessary for reproduction in one of the sexes, necessary for the particular habit of life of that sex, can still be developed further by sexual selection if higher levels in its development increase the chances of finding or keeping a mate. As Darwin implies, though natural selection must come first in determining the course of evolution, sexual selection may follow. Fisher (1930) fully appreciated the cogency and precision of Darwin's explanation of the theory of sexual selection. He advanced the theory further by explaining how female preferences for more highly adorned males could also evolve by both natural and sexual selection. This was certainly a missing link in the chain of Darwin's argument. Darwin could only assume that females would possess the necessary powers to discriminate between the more and the less highly adorned males. His critics, including Huxley (1938a, b), considered that Darwin was postulating the presence of an highly developed aesthetic sense in the females. They therefore rejected the idea of female choice as a mechanism of sexual selection. As we shall see, however, Huxley (1938a) in his discussion, 'The present standing of the theory of sexual selection', hopelessly confused the phenomena of natural selection, sexual selection and group selection in a review that was nevertheless influential in denying the importance of sexual selection in the evolution of male adornments as characters for sexual display. Darwin, with his customary clarity of thought, argued that sexual selection for male adornments depended, not on the females' necessarily possessing 'the same sense of beauty as a cultivated man', but on their being able to discriminate between the adornments of the males, which observation showed they were in fact capable of doing.

Darwin was always interested in how selective advantage was gained. He was able to propose specific models for sexual selection. As he said:

Our difficulty in regard to sexual selection lies in understanding how it is that the males which conquer other males, or those which prove the most attractive to the females, leave a greater number of offspring to inherit their superiority than their beaten and less attractive rivals.

This presents no difficulty if the species is polygynous. Darwin was aware of the close connection between polygyny and the degree of sexual dimorphism. In mammals, as Darwin shows by many examples, polygynous species often have much larger and stronger males possessing impressive tusks or horns for combat – a correlation of sexual dimorphism with polygynous breeding which is to be expected if the sexual differences evolved by sexual selection. But sexual dimorphism is found in many monogamous species. To explain how sexual selection would operate in monogamous species Darwin put forward what Fisher called 'a most subtle theory'. Darwin postulated that the females' nutritional condition at the start of the breeding season would determine both the time they were ready to breed, the number of eggs they produced and number of offspring they reared, the better nourished breeding earlier in the season and producing more offspring. Hence those males, who can best compete for the females or attract them, will mate with the first females to breed. They will produce more offspring than less competitive or attractive males who will mate later in the season with less successful females. Earliness of breeding time has been shown to correlate with breeding success in many non-passerine birds producing exactly the conditions that Darwin postulated for sexual selection. Computer simulations of Darwin's model (O'Donald, 1973a, 1974, 1976) show that sexual selection can give rise to large selective coefficients in favour of attractive males whom the females prefer to mate with. The computer simulations also show that selection will take place even when the earlier breeders are not the more successful: provided that late breeders are much less successful than average, unattractive males, who are left unmated until late in the season, will always suffer selective disadvantage. Models of sexual selection in monogamous birds are analysed in Chapter 7.

Darwin had no experimental evidence of the existence of female

preference for particular male characteristics. He relied on a comparative method to determine what characters might have evolved
by sexual selection:

There are many other structures and instincts which must have been
developed through sexual selection – such as the weapons of offence and
the means of defence of the males for fighting with and driving away their
rivals – their courage and pugnacity – their various ornaments – their
contrivances for producing vocal or instrumental music – and their glands
for emitting odours, most of these latter structures serving only to allure or
excite the female. It is clear that these characters are the result of sexual
and not of ordinary selection, since unarmed, unornamented or unattractive males would succeed equally well in the battle for life and in leaving a
numerous progeny, but for the presence of better endowed males. We may
infer that this would be the case, because the females, which are unarmed
and unornamented, are able to survive and procreate their kind.

This states one of the principles Darwin used to examine many
examples of sexual dimorphism and infer their evolution by sexual
selection. He also inferred that the males were the more modified
sex because the females more often resemble the young of their
species than do the males. In describing sexual dimorphism in
birds, he put forward a number of general rules:

I. When the adult male is more beautiful or conspicuous than the adult
female, the young of both sexes in their first plumage closely resemble
the adult female, as with the common fowl or peacock; or, as
occasionally occurs, they resemble her much more closely than they do
the adult male.
II. When the adult female is more conspicuous than the adult male, as
sometimes though rarely occurs, the young of both sexes in their first
plumage resemble the adult male.
III. When the adult male resembles the adult female, the young of both
sexes have a peculiar first plumage of their own, as with the robin.
IV. When the adult male resembles the adult female, the young of both
sexes in their first plumage resemble the adults, as with the kingfisher,
many parrots, crows, hedge-warblers.
V. When the adults of both sexes have a distinct winter and summer
plumage, whether or not the male differs from the female, the young
resemble the adults of both sexes in their winter dress, or much more
rarely in their summer dress, or they resemble the females.

The operation of rules I, II and V suggest that sexual selection may
have been in action, either on males (rule I), or as in a few species
such as the phalaropes on females (rule II) or on males or both
sexes (rule V). Sexual selection can certainly be inferred when the
character which differs between the sexes develops only in the

breeding season and is lost after mating has taken place, such as the red throat of the male stickleback. Recent research has completely confirmed Darwin's theory of the evolution of such characters. Semler (1971) set up aquaria in which a red and non-red male stickleback had built their nests and set up territories on either side of a glass partition. When the partition was removed and females were introduced, they showed strong preferences to lay their eggs in the nests of the red males. Semler proved that they were indeed responding to the red throat: non-red males with an artificial red throat painted on with lipstick or nail varnish gained the same advantage as the natural red-throated males. The females did indeed discriminate in favour of the 'more attractive' red males: as Darwin claimed, this demonstration is sufficient to explain the evolution of the red throat by sexual selection. The behavioural mechanism of the females' ability to discriminate – whether, as Darwin's critics claimed, by the implausible development of an aesthetic sense, or not – is irrelevant to the questions whether the females do discriminate and what advantage the males gain. Darwin, relevant as always, took male competition and female choice as facts to be observed and verified. Sexual selection is then a deduction from the facts, assuming heritable variation arises in the characters by which the males compete and which the females prefer.

1.2   Origin of sexual competition and mating preference

As we have seen in the previous section, sexual selection is a theory to explain how different sexual characteristics determine the chances of mating and how a selective advantage is thus gained. But Darwin was also interested in the evolutionary causes of male competition and female choice. He gave his explanation in terms very similar to those used by modern sociobiologists such as Trivers (1972). Trivers, while claiming that Darwin's treatment of sexual selection was 'sometimes confused', then puts forward as his fundamental proposition the view that 'What governs the operation of sexual selection is the relative parental investment of the sexes in their offspring.' The male, investing so much less per offspring than the female, has less interest in each offspring and more to gain competing with other males, especially if he can cuckold them so that they rear his offspring. Trivers' fundamental proposition had already been expressed by Darwin in the following words:

> The female has to expend much organic matter in the formation of her
> ova whereas the male expends much force in fierce contests with his
> rivals, in wandering about in search of the female, in exerting his voice,
> pouring out odoriferous secretions, etc.: and this expenditure is generally
> concentrated within a short period ... On the whole the expenditure of
> matter and force by the two sexes is probably nearly equal, though
> effected in very different ways and at different rates.

This seems to put Trivers' argument into a nutshell. And Darwin
was exactly right in his realization that the overall expenditure of
the two sexes in reproduction must be about equal. This idea was
the basis of Fisher's well-known theory of selection for equal sex
ratios (Fisher, 1930).

Darwin also partly anticipated Fisher's theory of the 'runaway
process' of sexual selection, for he observes that evolutionary
modification by natural selection must always have a limit if the
conditions of life remain constant, whereas sexual selection will
continue without limit until the injurious effects of extreme developments of certain characters bring sexual selection to a halt by
powerful adverse natural selection acting against further development. This is exactly how Fisher considered the 'runaway process'
would eventually come to a halt. Darwin never attempted to
explain why the females would continue to prefer to mate with the
more extremely developed males. To objections that female choice
would be fickle and give advantage to a succession of different
characters, Darwin could only assert the contradictory proposition.
Fisher's theory of the evolution of mating preferences overcomes
this difficulty.

In Fisher's words (1930):

> If instead of regarding the existence of sexual preference as a basic fact
> to be established only by direct observation, we consider that the tastes
> of organisms, like their organs and faculties, must be regarded as the
> products of evolutionary change, governed by the relative advantage
> which such tastes confer, it appears, as has been shown in a previous
> section, that occasions may not be infrequent when a sexual preference
> of a particular kind may confer a selective advantage, and there-
> fore become established in the species. Whenever appreciable differences
> exist in a species, which are in fact correlated with selective advantage,
> there will be a tendency to select also those individuals of the opposite
> sex which most clearly discriminate the difference to be observed, and
> which most decidedly prefer the more advantageous type ... In such
> cases the modification of the plumage character in the cock proceeds

under two selective influences:
(i) an initial advantage not due to sexual preference, which advantage may be quite inconsiderable in magnitude, and
(ii) an additional advantage conferred by female preference, which will be proportional to the intensity of this preference. The intensity of the preference will itself be increased by selection so long as the sons of hens exercising the preference most decidedly have any advantage over the sons of other hens, whether this is due to the first or to the second cause ...

The two characteristics affected by such a process, namely plumage development in the male, and sexual preference for such developments in the female, must thus advance together, and so long as the process is unchecked by severe counterselection, will advance with ever increasing speed. In the total absence of such checks, it is easy to see that the speed of development will be proportional to the development already attained, which will therefore increase with time exponentially, or in geometric progression.

Fisher has clearly explained how preferences will evolve for particular characteristics. His explanation can be stated in population genetic terms: the gene for the preference is selected because it is passed to the sons of the females with the preference. These sons, who will tend to inherit the preferred and advantageous character, will increase in frequency. Thus the gene for the preference will increase in frequency at the same time. The genes for the preference and the preferred character become associated in linkage disequilibrium. The selection of the one selects for the other, increasing the selective advantage in favour of both. Genetical models of this process (O'Donald, 1967, 1977c) show that a mating preference can be selected just as Fisher postulated, but the rate does not increase geometrically. Much depends on the genetics of the character and the initial gene frequencies in determining the level of linkage disequilibrium and hence the rate of selection that may be attained. For example, selection for a dominant character and its preference may be very slow, and not at all a runaway process accelerating at a geometric rate. I shall discuss Fisher's theory and its implications in Chapter 8. There I describe results of a detailed analysis of a general model of the evolution of mating preferences for different phenotypes or genotypes.

Fisher's theory certainly explains the establishment and persistence of mating preferences for particular male characteristics. What Darwin could only assert is thus shown to be a necessary part of evolution by sexual selection, completing further causal links in the chain of deductive inference which constitutes the theory.

Fisher thought he had also explained why females would prefer males with more and more extreme developments of the character that had provided the initial means of the females' discrimination. According to his theory, however, the mating preference would evolve to favour the initial stage of development of the character. The mating preference would certainly increase the initial selective advantage; but this does not seem to entail, as Fisher assumed, a greater advantage for more extreme stages of development. O'Donald (1977c) suggested that more extreme developments would act as 'supernormal stimuli' to the females. This behaviour can be explained by the evolution of a shift in the position of greatest response to stimuli. If so, the evolution of a mating preference would also entail selection for the more extreme stages of development and hence the 'runaway process' postulated by Darwin and Fisher. This shift of behavioural response, which I discuss in section 8.1, completes the chain of inference of the theory of sexual selection.

Since Darwin could not himself explain the origin of female preference, its existence became a basic premise of his theory: female preference was a fact, verifiable by observation, that would explain the evolution of particular cases of sexual dimorphism. Darwin's theory was incomplete, also, because he had no knowledge of genetics: Mendel's laws were still in limbo and would not be re-discovered until almost thirty years after the publication of *The Descent of Man*. Fisher's book *The Genetical Theory of Natural Selection*, written thirty years after the re-discovery of Mendelism, is largely an account of his research to put the theory of natural selection into the framework of Mendelian genetics. Fisher made no attempt, however, to put the theory of sexual selection into this framework: he describes the evolution of mating preferences in terms Darwin himself might have used. Fisher's non-genetical treatment of sexual selection would explain his mistake in assuming that the 'speed of development' of the preferred character would be 'proportional to the development attained, which will therefore increase with time exponentially, or in geometric progression'. In fact, as shown in Chapter 8, the rate of selection of the preference depends on the degree of association, or linkage disequilibrium, between the genes. This may increase at first but later usually declines; a mating preference for a dominant gene may never increase significantly at certain initial frequencies. A genetical model is thus essential for the valid deduction of evolutionary progress.

Many evolutionary models are synthetic and inductive and do not rely on a specific genetical model for analysis and deduction. Fisher's own contribution was often synthetic and inductive rather than deductive: thus he widened the scope of the theory of sexual selection to allow for evolutionary change in the mating preferences; he did not attempt to give the theory an analytical and deductive treatment, as he had done for the theory of natural selection. At the Sixth International Congress of Genetics, Fisher (1932) distinguished the synthetic and inductive method from the analytic and deductive in the following words:

The title chosen for our discussion is 'Contributions of genetics to the theory of evolution', and that these contributions are of two kinds, somewhat sharply contrasted, is well illustrated by comparing HALDANE's subject, 'Can evolution be explained in terms of present known genetical causes?' with the heading under which I chose to speak, 'The evolutionary modification of genetic phenomena'. My own address might equally well have been entitled, 'Can genetical phenomena be explained in terms of known evolutionary causes?' The one approach, as you perceive, is analytic and deductive. Genetic studies are regarded as revealing the mechanism connecting cause and effect, from a knowledge of which the workings of the machine can be deduced and the course of evolutionary change inferred. The other approach is inductive and statistical, genetics supplies the facts as to living things as they are now, facts which, like the living things in which they occur, have an evolutionary history and may be capable of an evolutionary explanation, facts which are not immutable laws of the workings of things but which might have been different had evolutionary history taken a different course.

Thus in Darwin's theory the mating preferences are facts from which the evolution of particular characters can be deduced: in Fisher's theory, they are a product of evolutionary history, undergoing evolutionary change during the progress of sexual selection. In this book, my own approach to the study of sexual selection starts from an analytic and deductive point of view using Darwin's and Fisher's theories as premises. In Chapters 4, 5, 6 and 7 I set up genetical models of sexual selection and deduce their evolutionary consequences, fitting data to the models where possible. However, in Chapter 8, I discuss how the mating preferences, described in terms of genetical models, might themselves have arisen and become established as part of the same evolutionary process: Fisher's synthetic theory is thus extended and then used deductively to determine the degree of preference that may evolve for different phenotypes or genotypes.

1.3   Eclipse of the theory of sexual selection

Although Fisher's theory of female preference explained the origin and evolution of some of the phenomena that Darwin took for granted, the theory of sexual selection remained in the partial eclipse where it had been since the early days of genetics. It stayed there because Fisher's book was mathematical and difficult and because of the publication of two papers by J. S. Huxley: 'The present standing of the theory of sexual selection' in *Evolution* edited by G. R. de Beer (Huxley, 1938a); 'Darwin's theory of sexual selection and the data subsumed by it, in the light of recent research' (Huxley, 1938b). The essence of Huxley's articles is to deny the importance of female preference in the evolution of sexual dimorphism, and characteristically, to replace the term sexual selection with terms of his own coinage. Female preference had indeed been denied since the first publication of Darwin's theory, on the grounds that it assumes that females have a highly developed aesthetic sense which they exercise in their choice of male. Most writers followed Wallace who considered that sexual dimorphism had evolved by natural selection for protective colouring of the females during incubation. But these theories were seldom examined in detail. Darwin's 'law of battle' was widely accepted – that males do fight for the females and will be selected for strength, pugnacity and the development of weapons; but female preference was rejected. Dewar and Finn (1909) were unusual in discussing sexual selection at length, but came to the same conclusions:

It is argued, with some show of reason, that it is absurd to credit birds with aesthetic tastes equal, if not superior, to those of the most refined and civilized of human beings.

And since

battle thus constantly decides the question of pairing, and in cases where, by hypothesis, the female should have most choice she has simply to yield to the victor.

Like some of the objections made against mimicry theory and natural selection, Dewar and Finn suppose that female choice must be either all or nothing: yet choice could still be exercised in favour of some of the victors and against others. They produce this fallacy in another of their objections to the theory:

Unfortunately for the theory of sexual selection, there is evidence to show that the cock Paradise Fly-catcher (*Terpsiphone paradisi*) in immature plumage is quite as successful in obtaining a mate as is the cock in his final plumage. The cock of this beautiful species has a chestnut plumage in his second year and a white one in the third year and subsequent years of his life. Nevertheless, a considerable proportion of the nests belong to chestnut cocks.

To Dewar and Finn, a 'considerable proportion' is the same as 'quite as successful as'. Unlike so many of his early critics, Darwin was always clear that selection was a relative and statistical phenomenon. To test for sexual selection, proportions of phenotypes must be known in the population of males from which the females will make their choice as well as the proportions among males who found mates.

Dewar and Finn also rejected Wallace's theory. This is refuted by observations of species in which the brightly coloured males, such as the cock Paradise Fly-catcher, incubate the eggs. As Darwin also pointed out, it is wholly implausible when applied to characters of sexual display, such as vocal apparatus, which are not based on plumage. Dewar and Finn conclude 'We have weighed each theory in the balance and found it wanting'. They fall back on the vacuous 'mutation theory', fashionable at that time (1909): 'Presently a mutation appears which is confined to the male alone; thus arises the phenomenon of sexual dimorphism'.

Huxley's two critical papers (Huxley, 1938a, b) were the first attempts to assess the theory of sexual selection in the light of behavioural research. They were widely quoted as an authoritative refutation of the theory of sexual selection by female preference. Yet I find them hopelessly confused: sexual selection is confused with natural selection, natural selection with group selection. As usual, Huxley wishes to replace existing terminology with his own. He proposes the term 'intra-sexual selection' for Darwin's sexual selection by male competition, and 'epigamic selection' for selection of primary and secondary sexual characters that promote the union of gametes. Epigamic selection thus includes two different mechanisms of selection which Darwin had separated: selection for characters that are used for mating, such as copulatory organs, and characters that increase reproductive effectiveness and success – a form of natural selection; and sexual selection for characters that influence the chances of finding and securing a mate.

Darwin, as we have seen, was quite consistent in the distinction he drew between sexual and natural selection. Huxley (1938b), however, commented obscurely:

> Darwin further failed to draw a general distinction between interspecific and intraspecific selection, although in sexual selection, as defined by him, he gave the first example of intraspecific selection promoting individual success without advantage to the type.

Darwin of course distinguished sexual selection from natural selection both of which operate within and not between species. The meaning of the second part of Huxley's sentence is beyond conjecture. Huxley's reference to 'interspecific selection' and references throughout his two articles to natural selection being 'of benefit to the species' suggest that he has confused natural selection with group selection and also sexual selection with natural selection. That this was indeed the source of confusion is shown by the following sentences (1938a):

> The facts recorded in the previous section indicate that display may often be of advantage to the species in promoting more effective reproduction. Any resultant selection will therefore come under the head of Natural Selection, not Sexual Selection in Darwin's sense.

Since one of the 'facts' was the cumulative excitement produced by mutual display in a colony of seabirds, it is clear that in this case Huxley considered that advantage did indeed accrue to the group; though in other cases, such as the synchronization of rhythms and the production of ova, the advantage is obviously to the individual.

No doubt many aspects of display are concerned with the process of mating after pairing has taken place. Therefore they will have evolved by natural selection, and not sexual selection in Darwin's sense. But Darwin had explained that characters developed primarily for the process of mating may also influence the chances of finding or keeping a mate and hence will evolve partly by sexual selection. As an example of this, Darwin suggested that if prehensile organs help to 'prevent the escape of the females before the arrival of other males, or when assaulted by them, these organs will have been perfected by sexual selection'.

Huxley discusses three main functions of display and threat actions:

(i) psycho-physiological effects of display and threat actions;

(ii) mutual display and 'mutual selection';
(iii) 'ritualization' of display and combat.

Psycho-physiological effects of display and threat actions

Evidence now exists to show that display may induce a psycho-physiological state of readiness to mate, irrespective of any possibility of choice (Drosophila – Sturtevant, 1915; newts – Finkler, 1923; etc.).

It is amazing that Huxley should quote here Sturtevant's experiments designed to show the operation of mating choice for mutants and wild-types in *Drosophila*.

In birds, display may synchronize male and female rhythms of sexual behaviour, initiate physiological changes leading to rapid growth of oocytes and eventual ovulation, while threat actions and display jointly may be necessary for reaching the threshold of effective reproductive behaviour ...
    Such effects, when positive, may be additive for members of a group, so that large colonies of socially nesting species may lay earlier and more synchronously than small colonies.
    These effects directly promote effective reproduction and need no special category of 'sexual selection' to explain their origin.

In this statement we can see the possibility of group selection being hinted at. However, we can certainly agree with Huxley that the origin of 'these effects' can be explained by natural selection. Darwin's and Fisher's point is that they can be further enhanced by sexual selection. Huxley himself suggests possible behavioural mechanisms for sexual selection, either by variation in the threshold of 'effective reproductive behaviour', or by variation in the intensity of the stimulation needed to reach the threshold. Both mechanisms would produce variation in the chances of finding and keeping a mate: if a male with a more effective display can more readily induce a female to mate, he may be less likely to lose her to another male before he has mated with her. The evolution of a mating preference by the Fisher process would give an increasing sexual advantage to the most effective males.

Mutual display and 'mutual selection'

Mutual display has the same functions as unilateral display in leading to more effective reproduction.

It also has the special function of acting as a bond to keep members of a pair together during the breeding season.

Here, too, sexual selection could enter if more than one clutch is produced in a season, so that males who can maintain a closer pair-bond are less likely to become cuckolds.

'Ritualization' of display and combat

Sham fights are to the advantage of both parties in avoiding combat. Huxley suggests that such fights, and sparring bouts like those of ruffs on their breeding grounds, raise the level of sexual excitement, and thus have a psycho-physiological function. If so, an advantage could only accrue to a group, since it can hardly be to the advantage of an individual male to stimulate another to mate.

In both his articles, Huxley refers to Fisher (1930). Of Fisher's theory, he says (Huxley 1938b):

> Fisher (1930, p. 131, seq.) on theoretical grounds believes that when there exists intra-sexual advantage among males, dependent on display, this and the capacity for the female to be stimulated by the display will reinforce each other, so that selection will initiate a 'runaway' evolutionary process which will proceed very rapidly to the utmost possible limit, sometimes leading to characters deleterious in the general struggle for existence.

In my quotation from Fisher's theory, it will be seen (section 1.2) that Fisher supposed that any advantageous character in the males would bring a concomitant advantage to those females who could recognize the advantageous males and would prefer to mate with them, thus selecting for a mating preference that would further increase the males' selective advantage. The character that provided the means of discriminating the advantageous males would thus evolve to become a display character as an object of female preference. Huxley appears to have misunderstood the basis of Fisher's theory, since an 'intra-sexual' advantage derived from male competition is not the starting point of Fisher's theory.

Yet in spite of the state of confusion in Huxley's ideas on sexual selection, his two articles were strangely influential. As recently as 1968, Lack, in his book *Ecological Adaptations for Breeding in Birds*, completely accepted Huxley's views. More recently, in a book of essays *Sexual Selection and the Descent of Man* edited by B. Campbell (1972), Huxley's articles are quoted with approval by the editor in his introduction:

> Julian Huxley was one of the first to discuss it [the theory of sexual selection] with authority in this century in his papers of 1914 and 1938.

Mayr (1972) and Selander (1972) in their essays in the same book also quote Huxley's articles with approval, though at the same time providing the evidence of female choice that refutes Huxley's point of view. Selander concludes with full support for Darwin:

The present review shows, I believe, that the theory of sexual selection is essentially correct as stated by Darwin – a conclusion reached earlier by Huxley (1938a, b) and Maynard Smith (1958).

And yet in fact the main purpose of Huxley's articles was to deny the importance of sexual selection by female choice in the evolution of characters for sexual display, and then to suggest mechanisms of natural selection, or group selection confused with natural selection, to explain their evolution. According to Huxley (1938a)

It is clear that Darwin's original contention will not hold. Many of the characters which he considered to owe their evolution to sexual selection do have value to the species in the general struggle for existence, and not merely in the struggle between males for reproduction.

Female choice for particular genotypes or phenotypes is now so well established that Selander may simply have been reading into Huxley's articles what he wanted to find there.

1.4   Reinstatement of the theory

With the publication of Campbell's *Sexual Selection and the Descent of Man* (1972), the reinstatement of Darwin's theory of sexual selection was complete. In this book, we find Ehrman's review of experiments on mating choice in relation to genotype and Selander's review of the evolution of sexual dimorphism in birds, both leaving the theory of sexual selection in the form that Darwin left it, but adding a great deal to the mechanisms of competition and choice.

Unfortunately, however, Ehrman introduced a confusion of terminology in her essay. She defined sexual selection as 'all mechanisms which cause deviations from panmixia'. On this definition, therefore, assortative mating would always be a case of sexual selection, and Ehrman does indeed discuss examples of assortative mating in man. Yet in a monogamous species, assortative mating does not necessarily produce a mating advantage, nor indeed any selection at all. While only if female mating preferences are expressed assortatively, in association with the preferred genotypes, will any deviation from the Hardy–Weinberg frequencies of random mating be produced when the population has reached genetic equilibrium. Sexual selection must therefore be distinguished from assortative mating (see section 1.5). Observations of assortative mating may, however, reveal the existence of mating choice, provided the assortment of genotypes or phenotypes is not a function of their separation into different ecological niches. Sexual selection has been reinstated as an important factor in the evolution of sexual differences largely because observation and experiment have revealed the widespread operation of mating choice, especially in insects and birds. No doubt it is equally widespread in mammals but probably dependent upon characters recognized through the highly developed non-visual senses of mammals. The many and very striking adaptations that we see in birds have evolved as sexual stimuli to be received by visual senses very similar to our own. If we had more highly developed senses of sound or smell, we should be more likely to appreciate sexual differences that have evolved as stimuli to these other senses.

Genetic differences provided the means of observing mating choices. Sturtevant (1915) did the first experiments, comparing frequencies of matings of different mutants and wild-types of *Drosophila*. He used eye and body colour mutants, and wing mutants, in his experiments, getting particularly large preferences in favour of red-eyed males when either red- or white-eyed females were given the choice of red- and white-eyed males. Experiments on similar lines were carried out by Merrell (1949, 1950), Reed and Reed (1950), Tan (1946) and Maynard Smith (1956), all showing the prevalence of female choice.

Petit (1954) made the important discovery that mating choice can be frequency-dependent: the number of matings per male achieved by males with different genotypes (e.g. wild-types and Bar eye) may be increased for those males with the less common genotype. This discovery was exploited and extended in Ehrman's and Spiess' many experiments on frequency-dependent mating success (Ehrman, 1967, 1968, 1969, 1972; Spiess, 1968; Ehrman and Spiess, 1969; Petit and Ehrman 1969; Spiess and Spiess, 1969). In these experiments, matings were observed in chambers in which males of two genotypes were present at some given ratio in a particular range of tested ratios of the genotypic frequencies. Thus in one of Spiess and Spiess' experiments, the frequencies of males homozygous for the inversions *AR* and *PP* and the frequencies with which they mated are shown in table 1.1. These data show

Table 1.1   *Numbers of matings by males at different genotypic ratios*

Number of males		Observed numbers of matings	
AR	PP	AR	PP
2	18	51	214
4	16	63	135
6	14	38	60
8	12	56	43
10	10	49	37
12	8	69	37
14	6	78	19
16	4	182	41
18	2	292	40

that *AR* males mate more often than *PP* males and their relative advantage is greatest at their lowest frequencies. At the frequencies 0.1 (2*AR*:18*PP*) and 0.2(4*AR*:16*PP*), an *AR* male mates about twice as often as a *PP* male. But at the frequency 0.8 (16*AR*:4*PP*) an *AR* male mates only slightly more often than a *PP* male, and at the frequency 0.9 (18*AR*:2*PP*), *PP* has a slight advantage. Ehrman and Spiess (1969) argued that these results showed that the expression of mating preferences was frequency-dependent. They supposed that females become habituated to the display of the more common males and then respond to the display of an unusual male because the state of habituation is broken. However O'Donald (1977a) showed that a model with constant preferences for each genotype would fit the data perfectly well. A constantly expressed preference for one or other of the genotypes will necessarily give rise to frequency-dependence in the males' mating advantage: if a constant proportion of females mate preferentially, then the males with the preferred genotypes must mate more often when they are rare than when they are common. Spiess and Ehrman's data can be fitted to several different models of the mode of expression of female preference. Each of these models assume that some females may express mating preferences while others always mate at random. The preferences may be expressed with different intensities according to the females' frequency of encounter with the males. I have analysed three models for the expression of female preference.

(i) In models with complete preferences (the 'C' models analysed in chapter 4), the females with preferences always mate with the males they prefer.

(ii) In models with partial preferences (the 'P' models analysed in Chapter 5), the females either mate preferentially if they meet one of the males they prefer within a certain number of chance encounters with courting males, or else they mate at random.

(iii) In models with frequency-dependent preferences (the 'PF' models analysed in Chapter 6), the females mate preferentially in proportion to the number of preferred males among the total number of males they encounter.

I have fitted each of these models to data of Spiess and Ehrman's experiments. Data of some of the experiments can be fitted satisfactorily to all of the models. Data of other experiments are more satisfactorily fitted by the partial or frequency-dependent models than by the complete model. (See section 6.9 for a comparison of the fit of the models.) Frequency-dependent male advantage has also been observed in a fish, the Guppy, *Poecilia reticulata* (Farr, 1977) and in a parasitic wasp, *Mormoniella vitripennis* (Grant, Snyder and Glessner, 1974). Such frequency-dependence should be a widespread phenomenon, since most of the models of female preference necessarily give rise to it. But it can only be demonstrated in an extensive series of experiments under carefully controlled conditions. Experiments at one particular frequency of the competing males can only give estimates of the overall proportion of females expressing a preference. For example, in Semler's experiments (1971) female sticklebacks were given the choice of laying their eggs in the nests of two males, one of the males having the typical re' throat of the male stickleback in breeding condition and the other being genetically non-red. His data on the mating advantage gained by the red males fit a model in which 43 per cent of females have exercised a preference for the red males. A constantly expressed mating preference of this magnitude will give rise to a very great advantage for the preferred character at low frequencies: at a frequency of 10 per cent, red males will be mating 8.5 times more often than non-red males. But, by the time the red males have reached a frequency of 90 per cent, they will be mating only 1.8 times more often than the non-red males.

It is much more difficult to find evidence of mating advantage in natural populations. The frequencies of matings between different phenotypes or genotypes may give evidence of mating choice if the preferences are assortative, for the frequencies will deviate from those of random mating. But in the absence of any such association between the preference and the preferred character, the overall frequencies of matings will be random. If the sequence of the matings is known, this may give evidence of preference, since males with preferred characters will tend to obtain mates before the others. In the Arctic Skua, when pairs are being formed, the dark plumaged, melanic phenotypes, which are common in the southern parts of the breeding range, mate sooner on average than the non-melanics (Berry and Davis, 1970; O'Donald, Wedd and Davis, 1974; O'Donald and Davis, 1977). Estimates of mating preferences can be computed to fit the different distributions of breeding times of the melanic and non-melanic phenotypes. At the same time Davis and O'Donald (1976a) obtained data on the total number of matings between the phenotypes in different colonies and found very significant assortative mating of the heterozygous, inter-mediate melanic phenotype. The estimate of the assortative mating preference agrees closely with the estimate from the breeding times. (Section 7.3 gives the details of these calculations.) However the preference for the homozygous, dark melanic was not assortative. Exactly as Darwin suggested in his theory of sexual selection in monogamous birds, the melanic male Arctic Skuas gain an advantage by breeding earlier in the season (sections 7.1, 7.4).

Eanes, Gaffney, Koehn and Simon (1977) used a different method of studying sexual selection. They were able to compare frequencies of allozymes of phosphoglucomutase and leucine naphthylamidase in mating pairs of the Four Spotted Milkweed Beetle *Tetraopes tetraophthalmus*, with the corresponding allozyme frequencies in the total population. This was possible because the beetles can be found both as solitary individuals and as copulating pairs. Since a number of separate populations with different frequencies of the allozymes could be sampled, frequency-dependence of mating ad-vantage could also be tested. The data showed statistically signi-ficant differences between some of the mating and non-mating samples. There was a strong suggestion of frequency-dependence in the relative mating advantage. This was not apparently statistically significant according to a rank correlation test, but a specific model of frequency-dependence was not fitted to test for the significance of this effect.

Extensive data on mechanisms of sexual selection in natural

populations are clearly going to be very difficult to obtain, even in species very favourable for the collection of the data. Data on assortative mating are relatively easy to collect if mating pairs can be observed; but, as we shall see next, assortative mating is not the same as, and may be much less common than, sexual selection.

1.5   Sexual selection and assortative mating

In theoretical discussions, sexual selection has sometimes been taken as synonymous with assortative mating. We have seen (section 1.4) that Ehrman (1972) defined sexual selection as any deviation from panmixia. Eanes, Gaffney, Koehn and Simon (1977) in the introduction to their paper on sexual selection in the milkweed beetle refer to a number of theoretical studies of sexual selection. These are all, however, references to the theory of assortative mating. None are references to the theory of sexual selection in which genotypes do not assort (O'Donald, 1962, 1963, 1967, 1973b, 1977c).

In O'Donald's original model (1960) of mixed assortative and random mating, it was assumed that the population was monogamous and all individuals found mates. Karlin and Scudo (1969), Scudo and Karlin (1969) and Karlin (1969) extended this model in a number of ways. In particular, they assumed that if all the assortative matings take place before any of the random matings, the later random matings might be less fertile. Karlin assumed this reduction of fertility was proportional to the frequency of the random matings, but gave no biological reasons for such an assumption. Darwin had provided the biological reason of reduced nutrition for a reduction of fertility of later matings in the breeding season. This is now well established for many non-passerine birds: reproductive success declines as the breeding season advances, the later-breeding birds producing smaller clutches and fledging their chicks with less success (see Chapter 7, sections 7.1 and 7.2). In O'Donald's models of this selection (O'Donald, 1973a, 1974, 1976 and sections 7.2, 7.3 of this book), the breeding season as a whole is divided up into successive intervals. In each interval a certain proportion of females arrive ready to breed. Some of the females exercise mating preferences and the others mate at random. Both preferential and random matings therefore take place in each interval. This must be biologically more realistic than assuming that all preferential matings occur in the earlier intervals and all random matings in the later intervals. Unfortunately the biological realism produces a complexity that rules out mathematical analysis: the models can only be analysed by computer simulation (section 7.4). Now in this model, sexual selection takes place regardless of whether the female preferences are assortative or not: females with preferences mating in the earlier intervals have more opportunity to exercise their choice and the preferred males gain an advantage from their earlier average mating time. Assortative mating in a monogamous species can therefore give rise to sexual selection exactly as defined by Darwin.

I have assumed in these models that the females with preferences will eventually mate with any male if no males of their preferred choice are left unmated. If these females were only to mate with the males they preferred, then there could be a reduction in fertility later: some later females might not find any males they preferred and might therefore remain unmated. This feature is contained in some assortative mating models of Karlin (1969) in which the random matings take place first, before any of the preferential matings.

In models with polygyny, the preferred males who mate preferentially remain available to mate at random: sexual selection in Darwin's sense always takes place with or without assortment in the matings. Karlin and O'Donald (1978) set up some models with polygyny in which parameters for mating preferences without assortment were combined with parameters for assortative mating preferences. Karlin and O'Donald described these models as 'models combining sexual selection with assortative mating', the 'sexual selection' referring to the selective effects of the non-assortative preferences. This distinction was useful in discussing the evolutionary processes occurring in the combined model but is a more restrictive use of the term 'sexual selection' compared to Darwin's usage. Darwin's own definition of sexual selection was perfectly clear and must be given the priority of history. A new term is therefore needed to describe the process of mating choice when assortative mating is excluded. 'Random preferential mating' would describe this process and thus distinguish it from 'assortative preferential mating'. Both types of preferential mating entail sexual selection in polygynous organisms. Sexual selection also occurs in monogamous organisms if later matings are less fertile than earlier ones: the preferred males then gain an advantage over other males in accordance with the model Darwin proposed.

In assortative mating, like usually mates with like: the preferred males are chosen by those females who also carry the genotypes that determine the preferred character. The assortment could also be negative: the preferred males would not then be chosen by females carrying genotypes for the preferred character; unlike matings would predominate. Negative assortment is very pronounced among matings of melanic and non-melanic pigeons (Murton, Westwood and Thearle, 1973; Murton and Westwood, 1977). But it does not seem to be a widespread phenomenon: many biological processes, such as divergence in overlapping niches and the evolution of isolating mechanisms, would only produce positive assortment.

In any form of assortative preferential mating, the expression of preference is restricted to females with particular genotypes at the loci which determine the preferred character. Random preferential matings are not restricted to females with particular genotypes at these loci. Random preferential mating should be more common, therefore, than assortative preferential mating: special mechanisms are not required to suppress the preference in some individuals but not in others. Random preferential mating is more difficult to detect, however. In monogamy, when preferences are expressed without assortment, the overall frequencies of matings are those of simple random mating (see table 3.1, Chapter 3). The existence of a mating preference cannot be detected without knowledge of the sequence of the matings. With assortment, deviations from random mating must arise and can be used to estimate the assortative mating preferences (Davis and O'Donald, 1976a).

Sexual selection with random preferential mating is the main subject of this book. Models of polygynous matings are analysed in Chapters 4, 5 and 6 and models of monogamous matings in Chapter 7. In Chapter 8, models of the origin and evolution of mating preferences are analysed numerically by computer simulation. These computer models show that preferences must often become associated to some extent with the preferred genotypes, thus giving rise to preferences that are partly assortative in expression. Combined models of preferences both with and without assortment (Karlin and O'Donald, 1978) are then described and fitted to data in Chapter 9.

2

Ecological and behavioural mechanisms of sexual selection

2.1 Potential for sexual selection

Why should sexual selection take place at all? This question and the corresponding question about natural selection have similar answers. The potential for reproduction and population increase is greater than the actual capacity. Sexual selection occurs because the males' high potential for fertilization is limited to the number of ova which the females actually produce and in which they invest much of their energy for reproduction. Female reproductive rate is a limiting factor in the intrinsic rate of increase of a population.

Darwin realized that the two sexes must invest an approximately equal effort in their different overall contributions to reproduction. This follows because their genetic contribution to the next generation is approximately equal. The males' investment in gametes is small, however; the females represent the limiting resource; and therefore the males must invest much of their effort in competing with each other. As a corollary, the males cannot afford to be choosy. Since one male can fertilize many females, a male who waits to give himself a choice may find himself without a mate. But a female can always be sure of fertilization in the end. She can wait to give herself the opportunity of choosing a better male who will produce fitter offspring. Male competition and female choice are the inevitable mechanisms of sexual selection.

Female reproductive rate is itself an ecological adaptation. It is selected for an optimum within the constraint that the net reproductive rate averaged over a generation can never differ by much from a value of 1.0. The net reproductive rate in a population, R_0, is defined by the formula

$$R_0 = \sum_x l_x b_x$$

where l_x and b_x are the survival and reproductive rates at age x. In a stable population we must have $R_0 = 1.0$; if $R_0 > 1.0$, the population is increasing in numbers; if $R_0 < 1.0$, it is a declining population. Suppose the survival rate increases in ameliorating conditions. Then the population will increase since the values of l_x will have been raised. This will increase the density-dependent pressure on reproduction: it will eventually become more difficult to rear offspring produced at the original rate b_x. As Lack has shown (Lack, 1954, 1966), the clutch size of birds is adjusted to an optimum at which the number of chicks hatched corresponds to the maximum number that can be reared successfully to fledging. If reproductive rates are higher than this optimum, the parents will be unable to provide for all the chicks they hatch and they will fledge fewer chicks than they would have done at an optimum, lower reproductive rate. As a population increases, so does competition for food, thus reducing the optimum reproductive rate. Selection for the new optimum rate increases the number of chicks fledged and hence the population size and density. As the population continues to increase, so the optimum rate declines and further selection for the new rate takes place. Eventually the actual rate and optimum rate coincide: values of b_x have then been reached at which stability has been attained and $R_0 = 1.0$. Similarly, if mortality has increased and the population is declining, then values of b_x are either raised by selection, eventually restoring the population to stability, or the population becomes extinct.

These evolutionary processes may give rise to alternative reproductive strategies: a species subject to high annual mortality may become an 'r' strategist with a high reproductive rate able to colonize a vacant ecological niche quickly at a high rate of increase, r. A species with low mortality may become a 'K' strategist maintaining a stable population at its carrying capacity, K.

Among the birds, seabirds have a generally low annual mortality – about 5 to 10 per cent in petrels (Procellariidae) and gulls (Laridae). Therefore their reproductive rate is correspondingly reduced. Passerines, however, may have an annual mortality of as much as 50 per cent. They must lay large clutches of eggs and often lay more than one clutch in the year. The females, by the size of the clutch they produce, determine the population size and growth. The demography of a population may therefore be described with reference to the females alone. Relative fitness of different phenotypes or genotypes is measured by the ratio of their rates of

increase over a period of one generation. Strictly, this, too, would
only apply to selection of females. But intrinsic rates of increase of
males can be found in a similar way by assuming that the males'
reproductive rates are given by their fertilization rates. The intrinsic
rates thus obtained can then be used to measure the selection
acting on males. The sexual component of selection will be a
function of differences in male fertilization rates caused by the
males competing for the females and the females choosing between
the males.

The disparity of the males' actual and potential fertilization rates
determines a potential, or opportunity, for sexual selection. The
actual intensity of sexual selection will always be much less than its
potential: the energy used by the males in competition with each
other is not unlimited; it must be less than the total amount of
energy spent on reproduction as a whole. Sexual selection is also
limited by the mating system: polygyny enables more of its potential to be realized than monogamy. In monogamous species, the
realization of the potential for sexual selection is limited to the
selective advantage that may be gained by finding a mate earlier
rather than later in the breeding season. In any mating system,
however, male competition and female choice determine the
sequence in which the matings take place. The sexual advantage
gained is then a function of the mating system and the species'
behavioural ecology during the breeding season.

Sexual selection is obviously more intense when matings are
polygynous. In successive polygyny (Murton and Westwood, 1977),
the male has successive broods with different females. Murton and
Westwood suggested that this behaviour may have evolved from
species with transient multiple pairing such as the Penduline Tit
Remiz pendulinus, the Wren *Troglodytes troglodytes* and the Pied
and Collared Flycatchers *Ficedula hypoleuca* and *F. albicollis*.
Successive polygyny is found in certain colonial or territorial
species, such as the Tricolored and Red-winged Blackbirds *Agelaius
tricolor* and *A. phoeniceus*. The male Red-wing holds a permanent
territory and at successive intervals during the breeding season
mates with four or more females. Less successful males with poor
territories or no territory at all remain as surplus non-breeding
males. The successful males can therefore gain a large advantage.

Murton and Westwood (1977) use the term simultaneous polygyny for what they regard as a further evolutionary development
of the mating system: the male holding the territory mates with

several females during the same period rather than successively. This behaviour will lead to an even greater realization of the potential for sexual selection. According to Murton and Westwood.

When the male's main function is to fetilize a female, after first acquiring a territory, conditions exist for intense sexual selection and the development of brilliant plumage and elaborate displays, as in the birds-of-paradise (Paradisaeidae), or complex display structures as in the bower birds (Ptilinorhynchidae). Most species perform in their own individual territories but some resort to a communal display ground or lek and hold small 'courts' which they defend within the main area. Here there can be direct competition for the female's attention. Among the nidicolous birds which have leks are two species of birds-of-paradise, most manakins (Pipridae), the Cock-of-the-Rock *Rupicola rupicola* (Cotingidae), Jackson's Widow-bird *Euplectes jacksoni* (Ploceidae) and some humming-birds (Trochilidae). Among nidifugous species, lekking behaviour is recorded for the Great Bustard *Otis tarda* and is especially evident in the grouse (Tetronidae) and waders (Charadriiformes).

In primates, sexual dimorphism in size depends on polygyny. Clutton-Brock, Harvey and Rudder (1977) found that the degree of sexual dimorphism increased significantly with an increasing sex ratio of females to males in breeding groups. Monogamous species show little or no dimorphism; polygynous species are often highly dimorphic with the males in some species twice the size of the females. The mating system itself, whether it be monogamous, promiscuous or polygynous, or much more rarely polyandrous, is presumably in origin an evolutionary adaptation to the requirement of rearing the offspring in relation to the supply of food. In species with a short breeding season, both parents may be required to take an equal share in the feeding of the young. This will obviously maintain the advantage of a state of monogamy: the male invests as heavily as the female in the energy expended on parental care and he will have proportionately less energy to compete with other males. In birds that breed in the Arctic tundra where food is short after mid-summer, reversal of sex roles may even evolve, as in the Grey and Red-necked Phalaropes *Phalaropus fulicarius* and *P. lobatus*. In these species, the female, which is the more brightly coloured sex, defends a territory against other females and courts the males, while incubation and parental care are solely in the charge of the male. This is one of those rare cases, mentioned by Darwin, in which sexual selection has acted upon the female. Presumably the female contributes all her investment in reproduction to courtship and the production and laying of the eggs. She competes to be fertilized in time to lay her eggs so that they hatch just as food becomes available at the beginning of the very short breeding season. She is sexually selected as a result of the disadvantage incurred by breeding towards the end of the season when the food supply is running out. The Dotterel *Eudromias morinellus*, which breeds in the Arctic tundra and the alpine zone of the Scottish mountains, also shows a trend towards sex reversal similar to but not so pronounced as that of the phalaropes. Murton and Westwood (1977) trace the evolution of sex reversal in those species of waders (Scolopacidae) that have similar Arctic breeding distributions.

Monogamous mating may have a general advantage, as Murton and Westwood suggest, of synchronizing the breeding of long-established pairs. The breeding of established pairs of Arctic Skuas is synchronized to within about a day in successive years. But monogamy will tend to evolve towards polygyny if a female can easily produce her eggs and also raise her young without the male's help. If the males gain more advantage by competing for females rather than caring for their own young, then we may expect that first successive polygyny and then simultaneous polygyny will evolve.

Sexual selection, as we have seen, depends on the effects of male competition and female choice which determine the sequence of matings of those individuals with special characters for competitive ability and sexual display. The breeding ecology and mating system then determine to what extent the sexually favoured individuals mate more successfully or more often than the others. The mating system may also limit the sequence in which the matings can take place. In monogamous mating, the males, once they have mated, do not mate again: they are sampled without replacement by the females who, if they breed towards the end of the season, must make their choice from a more restricted group of males. In a large population, however, in the absence of assortment, the overall frequencies of matings are equal to the frequencies of random mating (see Chapter 3, section 3.1) and selection only takes place if matings with the preferred or more competitive individuals are the more fertile.

2.2   Darwin's 'Law of Battle'

Specific behavioural mechanisms have evolved by which the males compete with each other and the females select the males of their choice. Fighting is perhaps the simplest and most direct form of competition. Since Darwin first put forward the idea of sexual selection, it has generally been accepted that sexual selection takes place when rival males fight for the possession of a harem of females and that the selection will 'add to the size, strength and courage of the males, or to improve their weapons'. No doubt the particular details of male competition will be different in different species. The size and security of a harem may vary not only between species but also with the density of individuals in a colony and the frequency of males whose characteristics enable them to achieve dominance. Male competitive behaviour may lead to some very complex processes of selection. Competition for females between male dung flies *Scatophaga stercoraria* occurs in several stages each with interacting effects (Parker, 1970, 1974a, b).

Male dung flies search for females on the surface of fresh droppings of dung. The females arrive to lay their eggs on the dung. They are greatly outnumbered by the males who attack or strike other males in their encounters while searching for the females. After finding a female and mating, a male stays paired with her in a passive phase while she oviposits. But ovipositing and copulating females may be taken over by other males at frequencies of 32 per cent and 4 per cent respectively. Parker (1970) was able to show that the last male to mate with a particular female will fertilize an estimated 81.4 per cent of her next batch of eggs. Males who take over a female will therefore fertilize a large proportion of the eggs in her next batch unless taken over themselves. Parker calculates the overall advantage in terms of numbers of fertilizations that may be gained either by mating with virgin or non-virgin females, or by taking over ovipositing or copulating females. The selective advantages are complicated functions of the probabilities of taking over females and the proportions of eggs that may then be fertilized. Parker shows that the proportion of eggs fertilized also depends on the duration of copulation. The overall selection is a function of the males' searching behaviour, which is strongly density-dependent (Parker, 1974a), and their mating behaviour, which determines the number of eggs a male may be able to fertilize. A change in one aspect of behaviour will affect the consequences of another. For example, if a gene increased the duration of copulation, this would increase the proportion of eggs likely to be fertilized in subsequent batches and consequently alter the number of fertilizations achieved by matings with virgin or non-virgin females or matings after take-overs. Similarly, a different set of frequencies of these mating types would alter the advantage gained by copulating for a particular length of time. Genetic variations in the different aspects of the males' competitive behaviour thus interact in the determination of the final selective outcome. No simple model, applicable to competition in different species, could be constructed from this complicated set of interactions.

D. and B. Charlesworth (1975) proposed the simplest possible model of male competition in which males of one phenotype are k times more likely to find a mate than males of another phenotype. As shown in Chapter 3, sections 3.5 and 3.6, this model produces a consistent advantage of the better competitor with no frequency-dependence. It might apply to species in which some males are more likely to be able to collect and defend a harem. Even so, the behaviour in searching for females will usually be density-dependent, and hence frequency-dependent. Clearly, the Charles-worths' model does not fit the sexual selection of male dung flies whose different matings produce different numbers of fertilizations.

2.3 Territoriality and display

Males may compete for females in more subtle ways than merely by fighting for them. They may compete in the size of the territories they hold and by the intensity of their display and courtship. Of course the territory may have had to be fought for in the first place, but display and threatening behaviour is often sufficient to hold it. Perhaps the lark 'at break of day arising from sullen earth' does 'sing hymns at heaven's gate'; but the hymns are for victory in battle and they are answered without the need for actual combat. Since breeding grounds are areas of limited space, possession of a territory will exclude some other males from mating. A male who can hold a larger territory increases his chances of encounter with the arriving females who will more often land on a larger than on a smaller territory. Territory size varies widely in the Arctic Skua *Stercorarius parasiticus* (Davis and O'Donald, 1976b). Males with larger territories mate earlier in the breeding season. The relation between territory size and breeding date in newly paired birds is very close to what we should expect if females arrive successively and land at random on the males' territories (see figure 7.1, section 7.5). Since females do not in fact distribute themselves at random over the breeding grounds, however, another explanation must be found for the relation of territory size to breeding date. The Arctic Skua is polymorphic for a non-melanic and two melanic plumage phenotypes. Since mating preferences favour the melanics (see Chapter 7, sections 7.2 and 7.3 for details), the non-melanics, which also have the smaller territories, find their mates later than average in the breeding season. This may explain why birds with small territories breed late in the season. Sexual selection for melanism would thus indirectly select for increased territory size.

The Arctic Skua seems to be territorial both for mating and for rearing the nidifugous chicks who are allowed to wander freely within the confines of their parents' territories. Unlike the territories of many passerine birds, those of the Arctic Skua can have no function as a place to feed: Arctic Skuas feed almost entirely at sea by piracy, chasing other seabirds and forcing them to disgorge their food. Even in species in which the primary function of territory is to provide a safe place to feed, sexual selection may still be a subsidiary factor in the evolution of territoriality. We may expect this to be so whenever territorial behaviour restricts the opportunities for successful mating, either by prohibiting the mating of males without territories, or by reducing the chances of mating at the best time in the breeding season. Sexual selection will increase the size of territory above that required solely as a place to feed: the best size for one function will obviously not be the best for another.

The frequency and intensity of the males' display towards the females is another factor on which the chances of mating depends. It is often closely correlated with the threatening displays that are used in territorial defence. Usually a female is reluctant to mate when first approached and courted by a male: she is passive and discriminating, while he is eager and undiscriminating (Bateman, 1948). In *Drosophila* species, according to Ehrman and Spiess (1969),

There is no question that females determine mating frequency. Their receptivity varies considerably, but wariness on the part of these species' females [i.e. *Drosophila pseudoobscura* and *D. persimilis*] is evident, especially in slower-mating strains, though even in fast-mating strains of *D. persimilis* the first reaction of the female is reluctance to mate, and she will need more than one encounter with a male before accepting him.

But having encountered courting males and received stimulation up to the level of a particular threshold, further courtship then releases the females' mating behaviour. Males whose courtship is more vigorous and intense produce more sexual stimuli: they will more rapidly raise a female's level of stimulation, and thus elicit her response. The nature of the stimuli will of course differ widely in different groups of animals. Visual stimuli are likely to be especially important to birds, the stimuli being supplied by the particular colours and patterns of the males' plumage. In fish, too, visual stimuli determine mating response: the red throat of the male stickleback greatly increases his chance of persuading a female to lay her eggs in his nest (Semler, 1971; a brief description of Semler's experiments has already been given, Chapter 1, section 1.1).

In *Drosophila*, pheromonal, auditory and visual stimuli have all been shown to be components of mating behaviour (Bastock and Manning, 1955; Burnet and Connolly, 1973; Miller, Goldstein and Patty, 1975; Averhoff and Richardson, 1976). Ehrman and Spiess (Ehrman, 1969; Ehrman and Spiess, 1969) had shown that olfactory cues provided by males who were not taking part in mating could still influence the expression of female preference. Following up this demonstration, Averhoff and Richardson (1976) then obtained evidence of genetic variation in the pheromones that stimulate a female to accept a courting male. Volatile and non-volatile components of the pheromonal system were found to be determined by loci on different chromosomes. According to Averhoff and Richardson's model, a female pheromone first stimulates the males to courtship. The male then produces two pheromones, a non-volatile or only slightly volatile pheromone, which requires the presence of the males for transmission, and a volatile pheromone which can be passed through an olfactometer. Within certain inbred lines, however, females reject male courtship; presumably the effective pheromones were not being produced. By extracting chromosomes of individuals crossed to 'balancer–crossover–suppressor' strains, Averhoff and Richardson showed that the volatile pheromone was associated with the second and third chromosomes, and the non-volatile pheromone with the X chromosome and the fourth chromosome. Genetic variation in pheromones will obviously give rise to sexual selection in favour of those males that produce the more effective pheromone. Females will also express a preference between males if they differ in their receptivity to the different pheromones.

In birds, the male sex hormone, testosterone, largely determines the expression of sexual behaviour and display (Witschi, 1961). This is true even when the female is the active sex in courtship and display. In the example of the Grey and Red-necked Phalaropes already mentioned in section 2.1 of this chapter, we have seen that the females defend the territories and court the males by displaying their more brightly coloured plumage. In both sexes, the development of nuptial plumage is induced by the injection of testosterone propionate (Johns, 1964). The ovary of the female phalarope has a higher content of testosterone than the male testis. The female's brighter plumage is therefore a function of her higher testosterone level.

Higher levels of testosterone also increase territory size. Watson (1970) found that in the Red Grouse *Lagopus scoticus scoticus*, implants of testosterone increase both the aggressive components of behaviour and the territory size of the males. But level of testosterone is not the only determinant of territory size. Watson and Moss (1971) found that the territory size also varied inversely with nutritional condition. Birds which attain a higher nutritional state in the summer are then disposed to take smaller territories in the autumn and winter; the higher the nitrogen content of the heather, the smaller the territory. These observations coincide with what we should expect according to the theory that territory is primarily a place to feed: the higher the nutritional level of the birds and the better the quality of food available, the smaller the territory needs to be. But as I have already argued, territory may have several functions. After territoriality has evolved, primarily perhaps as an adaptation to secure a place to feed, it may then evolve further by sexual selection. If a male's sexual display and courtship, his territory size and his aggressiveness towards other males are all partly functions of his testosterone level, then he who is the victor according to the 'law of battle' will also possess the largest territory and will most attract the females. Sexual selection will act in his favour in all these aspects of behaviour. This argument might seem to imply that sexual selection would always increase the level of male testosterone. But as both Darwin (1871) and Fisher (1930) pointed out, sexual selection can never continue indefinitely without producing strongly deleterious effects. For a while, indeed, it may seem to be the 'runaway process' that Fisher postulated; but eventually it must become increasingly harmful to devote ever more energy to reproductive behaviour of various sorts: the red throat of the male stickleback is a stimulus both to the females and to the stickleback's predators. These considerations are clearly illustrated by Murton's work on the reproductive physiology, mating behaviour and ecology of the melanic and non-melanic phenotypes of the feral pigeon.

Murton, Westwood and Thearle (1973) studied a polymorphic population of feral pigeons in the Salford Docks in Manchester. They compared the breeding of this urban population with a population on the coast at Flamborough Head. The urban population contained a far higher frequency of the melanic phenotypes 'blue-checker' (C), 'dark blue-checker' or 'T-pattern' (C^T) and 'spread pattern' or 'velvet' (S): only 21 per cent are wild-type ($+$) in Salford; 70 per cent are wild-type at Flamborough. Genetically C^T is dominant to C which is dominant to wild-type; S is an allele at a different locus to the C locus, also dominant to wild-type. Murton, Westwood and Thearle observed a total of 195 matings between the four main phenotypes in Salford. They found strong negative assortative mating. Davis and O'Donald (1976b) fitted the data to a model of negative assortment in which females might prefer either wild-type, blue-checker or dark blue-checker, but only express their preferences if they do not themselves possess the phenotypes they prefer. At maximum likelihood, the model shows that 51 per cent of females preferred blue-checkers and 38 per cent preferred dark blue-checkers. Murton and Westwood (1977) should be consulted for the details of the model and the fit of the parameters; these are given in Appendix 5 of their book. The model certainly fits the data very well. The estimates of both mating preferences are highly significant: $\chi_1^2 = 24.013$ for the preference for blue-checker: $\chi_1^2 = 17.448$ for the preference for dark blue-checker. A non-significant value of $\chi_4^2 = 6.565$ is left for the residual heterogeneity after the model has been fitted.

The preferences in favour of the melanic morphs are correlated with differences in their reproductive physiology. Figure 2.1, taken from Murton, Westwood and Thearle's paper (1973) shows that the melanic morphs spread pattern and dark blue-checker (T-pattern) both develop larger testes than the other morphs. Spread pattern morphs show a much reduced level of testicular regression and also a difference from the others in the phase of their annual gonadal cycle. After birds have been kept in 24-hour cycles with only 8 hours of daylight, a smaller proportion of all melanic morphs regress into a non-breeding condition. In the population at Salford the melanic birds continue to breed throughout the year. The dark blue-checkers were found to have very significantly higher numbers

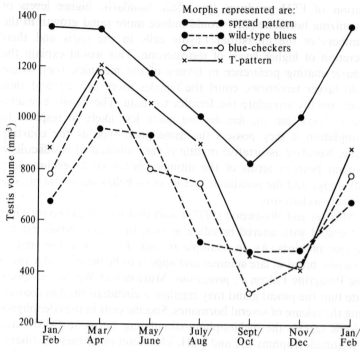

Figure 2.1   Mean testicular volume of various feral pigeon phenotypes at Salford, Manchester. Using a t-test the following differences were obtained: Sept./Oct., no significant differences; Nov./Dec., no significant differences; Jan./Feb., spread significantly larger than wild-type with $t_{34} = 3.72$, $P < 0.001$; Mar./Apr., spread significantly larger than wild-type with $t_{47} = 3.10$, $P < 0.01$; T-pattern significantly larger than wild-type with $t_{113} = 2.60$, $P < 0.01$. (Reproduced from Murton, Westwood and Thearle, 1973.)

of spermatocytes and spermatozoa per section of testis tubule than the other morphs. Thus to summarize these results, the spread pattern melanic develops the largest testis and shows less regression of testis volume in the autumn and winter. The dark blue-checker produces the most germ cells. Mating preferences favour melanics generally. These differences must be caused by differences in the capacity to secrete gonadotrophin and can be explained by a lowering of the threshold of photoresponse. A difference in photo-period with earlier secretion of the follicle-stimulating hormone (FSH) would certainly explain the higher counts of spermatocytes and spermatozoa in the dark blue-checkers, since the administration of FSH produces this effect. Similarly, higher levels of luteinizing hormone (LH) would induce more rapid growth in the numbers of the interstitial Leydig cells in the testis and their secretion of higher levels of testosterone. This would explain the greater mating preference in favour of the melanics: they would hold larger territories, court the females more actively and thus more readily stimulate the females to mate. The negative assortment shows that the females are much less likely to respond to stimulation if they possess the same phenotype as the courting male. Negative assortative mating is very unusual. It is difficult to explain both in terms of the ultimate factor of its evolutionary advantage and the proximate factors of its behavioural and physiological mechanisms.

Murton and Westwood (1977) noted that melanism is commonly associated with altered breeding seasons in several closely related species of birds. As an example of this, *Falco eleonorae* and *F. concolor* breed in late summer and appear to be melanic relatives of the Peregrine Falcon *F. peregrinus*. Murton and Westwood speculate that the pineal gland may regulate a circadian rhythm controlling the release of several hormones. Specific cells in the adenohypophysis secrete the melanophore-stimulating hormone (MSH) and the gonadotrophins LH and FSH. The secretion of these hormones takes place under the control of releasing factors secreted in the hypothalamus. If the releasing factors share the same photoresponse, a gene that reduced the photosensitive threshold would give rise to an earlier release of LH, FSH and MSH at higher overall levels. Melanism would thus be the pleiotropic effect of a gene that primarily adapts an individual to a longer breeding season and thus gives increased reproductive success in favourable environments. The melanic phenotypes of both pigeons and Arctic Skuas rear a significantly greater number of chicks than the non-melanics. But like characters for sexual display, a longer breeding period and increased reproductive success must eventually bring corresponding disadvantages. Melanic pigeons are characteristic of urban populations. In Salford, where Murton's study was carried out, spillage from storage and provender mills provides a continuous and unlimited supply of grain throughout the year. This can presumably supply the energy both to moult and at the same time to continue breeding. Where high-energy food is not available in unlimited quantities, the demand on a bird's energy budget is too great to allow breeding to continue into the autumn and winter while the moult is in progress. (For a discussion of the energy budget of birds in relation to the demands of reproduction and moulting, see Murton and Westwood, 1977, Chapter 7.) The longer breeding season would have more of a deleterious effect away from urban areas where birds would certainly be unable to find enough food to breed and moult at the same time. If melanics are maintained by a balance between sexual selection acting in their favour and natural selection acting against them (O'Donald, 1977c; Chapters 4 and 5 of this book) we should expect to find polymorphisms with higher frequencies of melanics in urban areas such as Salford: where food is in constant and unlimited supply, more energy can be found to spend on reproduction without necessarily reducing the energy needed to ensure survival.

2.4   Female choice

Females can be seen to respond more readily to some males than to others. Female sticklebacks prefer to lay their eggs in the nests of males with red throats (Semler, 1971). And this is true when other things are equal, for the females show just the same preference for non-red males who have been painted with an artificial red throat as they do for the natural red-throated males. The red throat acts as a signal which lowers the females' threshold of response. But only by an experiment like Semler's can it be proved that the discrimination in favour of the red males depends on differences in threshold in the females: as we have already demonstrated (section 2.3), in birds development of plumage for display, intensity of courtship, size of territory and aggressiveness are all functions of testosterone level. The male with the brighter nuptial plumage may be the more successful, not because his colours are more attractive to the females, but because he courts them more actively or persistently. And differences in level and intensity of courtship will produce similar effects to differences of preference within the females.

Individual females vary in their receptivity as Ehrman and Spiess (1969) have observed. Some females have lower thresholds than others. Suppose the females with the low thresholds can be stimulated to mate by the courtship of any male they encounter. Obviously they will mate with the males at random. Others with high thresholds only mate if they encounter a male with a sufficiently intense or persistent courtship to produce the extra stimulation they need. These females will exercise a mating preference for the sexually more active males who may also be the more brightly coloured. The effect will be the same as if some females had a lower threshold to the more brightly coloured males. In experiments in mating chambers on the frequencies with which females choose between different males, these different causes of the expression of a preference cannot be distinguished.

A similar result follows if females need the stimulation of a number of successive encounters with courting males before mating. In their previous encounters, some females will have been stimulated to a level close to their theshold; others will not. Those who are sufficiently close to their threshold will respond to any courting male at the next encounter. Others, not so close, will need further encounters, or an encounter with a sexually more active male before mating; hence they exercise a preference for the more active males. In section 3.2, I describe models of the frequencies of matings which depend on the probabilities of encounter with the different male phenotypes. These models are analysed in genetic terms in Chapter 5.

The same models will also describe the effects of differences in the females' thresholds of response towards different males. Suppose a female has a lower threshold of response towards males with a particular plumage phenotype and mates on encounter with any male of this type. However, she needs more encounters with other phenotypes before her level of stimulation has reached the higher threshold at which she will accept one of the others. It may be that some male characteristics act as a necessary signal that a female must receive before she can respond. Without it, males would find an infinitely high threshold against them. Females who required this signal would always exercise a preference in favour of the males who gave it. Such behaviour in a proportion of the females would give rise to the models with complete expression of preference. These models are introduced in section 3.1 and analysed genetically in Chapter 4.

In fact, mating behaviour must almost always be a response to several characters: some that are necessary before any further stimulation can be received; others that give a succession of stimuli and elicit a response when the level of stimulation has reached a particular threshold. This may be a possible mode of action of the pheromones of *Drosophila melanogaster* studied by Averhoff and Richardson (1976). The non-volatile pheromone would provide the essential stimulus. The volatile pheromone could then act on the receptor cells and at a certain threshold in concentration the females would respond. As Averhoff and Richardson point out, this system could theoretically give rise to frequency-dependence of response. If some males produced a variant of the volatile phero-mone which differed biochemically from that produced by the other males, then amounts of the two pheromones diffusing through an area would be proportional to the frequencies of the males. Averhoff and Richardson suggest that the receptor cells might accommodate to high concentrations of pheromone giving no response. If some females were more receptive to one of the pheromones, they would show a reduced frequency of response when males who produced this pheromone were common. There might equally well be a lower threshold, however, below which the females do not respond. Increasing concentrations of pheromone would then increase the frequency of response. These effects suggest the possibility of frequency-dependence in the expression of mating preferences. Sections 3.3 and 3.4 describe models with frequency-dependent preferences. They are analysed genetically in Chapter 6. They are all models in which females express preferences for particular phenotypes or genotypes of males. They are not models in which females accept any unusual male simply because he is rare and breaks their habituation to commoner males. This suggestion, which Ehrman and Spiess (1969) put forward to explain the mating advantage so often gained by males who are in the minority, is close to being the tautology that females choose rare males because they are rare. It cannot be quantified as a model that might fit some data and be refuted by others. If it were general for females to choose rare males simply because they were rare, the females would certainly find themselves with some highly undesirable partners. Such mating behaviour would obviously be mal-adaptive (O'Donald, 1978b). I shall demonstrate (Chapters 4 and 5, sections 4.5 and 5.5) that models of completely or partially expressed preferences can be fitted satisfactorily to Ehrman and Spiess' data on mating frequencies of *Drosophila*. Their data do not therefore entail a frequency-dependence of the expression of preference, even though models of frequency-dependent preferences do fit some of the data. When preferences for two different phenotypes are about equal, both complete and frequency-dependent expression of pre-ference give rise to a similar advantage of being the rare male: data that fit one model will also fit the other; neither model can then be

refuted simply by data on the frequencies of the matings of the rare
and the common males. Table 6.2 in Chapter 6, section 6.9, shows
how models of complete, partial and frequency-dependent preference compare in their goodness of fit to the same data taken
from several of Ehrman's and Spiess' experiments.

3

Models of sexual selection

3.1 Complete mating preferences

In Chapter 2 I showed that variation either in female response or in male courtship may give rise to sexual selection: a male's chance of getting a mate may be increased if he possesses a particular phenotype and reduced if he does not. In the simplest model, some females can respond only to the courtship of males that possess a particular character of sexual display. The red throat of the male stickleback may be an example of such a character. If some females never laid their eggs in the nests of the non-red males so long as red males had nests to lay in, then the red males would gain a selective advantage that depended on the proportion of females exercising their preference in favour of red males: the red males would always mate with the females with the preference and also mate at random with other females who did not discriminate between red and non-red males. This mating behaviour (discussed in section 2.4) would arise if the red throat were a necessary component for eliciting the response of some of the females. I call models based on this simple assumption 'models with complete preferences', or 'C' models.

Most of the models of sexual selection and assortative mating which have been described and analysed in the literature (O'Donald, 1960, 1962, 1967, 1973b, 1977a, 1977c; Karlin, 1969; Karlin and Scudo, 1969; Scudo and Karlin, 1969) are derived implicitly or explicitly from models with complete preferences. In all these models, certain proportions of females are assumed to have preferences for different male phenotypes or genotypes: they only mate with the males they prefer. If these female preferences are expressed only in association with particular genotypes, then the matings will be assortative. As defined in section 1.5, positive assortment occurs when females express preferences only if they too possess the genotypes that determine the preferred characters:

negative assortment occurs when a preference is expressed only in females who do not possess these genotypes. Table 3.1 shows the different matings that may arise with no assortment and with either positive or negative assortment.

The table also shows the difference between polygynous and monogamous mating systems. In polygyny, the males remain available to mate again after previously having mated: their frequencies stay the same for all preferential and random matings that take place. In monogamy, a male who has once mated is then removed from the pool of available mates: no selection for change in gene frequency takes place unless preferential and random matings differ in average fertility or some individuals remain unmated. In monogamous birds, as Darwin originally suggested (Chapter 1, section 1.1), earlier matings are more successful. Males who obtain their mates earlier in the breeding season gain a considerable reproductive advantage. As the females come into breeding condition, they choose their mates from the males who remain unmated in their territories. Since the earlier females have more males to choose from, the preferred males tend to get chosen as mates earlier in the breeding season. They thus gain a selective advantage.

Although sexual selection can thus operate in monogamous species, it will usually be much more effective with polygyny. The inferior males will probably not mate at all compared to the multiple polygynous matings of those who possess the desirable characteristics. Suppose two male phenotypes A and B occur with frequencies x and y. If there is a proportion α of the females who only wish to mate with A males and γ who only wish to mate with B males, then the matings with these males will take place at frequencies

$$\left. \begin{aligned} P_T(A) &= \alpha + x(1-\alpha-\gamma) \\ P_T(B) &= \gamma + y(1-\alpha-\gamma) \end{aligned} \right\} \qquad (3.1.1)$$

where the remaining $1-\alpha-\gamma$ of the females are assumed to mate at random. Since the females may also be phenotypically either A or B, these frequencies then determine frequencies of polygynous matings without assortment as shown in table 3.1. The model implies polygyny, since females with preferences always mate with the males they prefer even though preferred males may occur at a lower frequency than the females that prefer them.

The frequencies of matings with different males may be observed

Table 3.1 *Frequencies of matings with and without assortment of phenotypes*

(i) Frequencies of preferential matings alone.

α and γ are proportions of females with complete mating preferences in favour of phenotypes A and B. A and B have population frequencies u and v in both sexes ($u + v = 1$).

Female phenotypes	Frequencies of phenotypes of males mated preferentially					
	Without assortment		With positive assortment		With negative assortment	
	A	B	A	B	A	B
A	αu	γu	αu	—	—	γu
B	αv	γv	—	γv	αv	—

(ii) Total frequencies of both preferential and random matings.

Assortment and female phenotypes		Frequencies of phenotypes of males mated polygynously	
		A	B
Without assortment	A	$\alpha u + u^2(1-\alpha-\gamma)$	$\gamma u + uv(1-\alpha-\gamma)$
	B	$\alpha v + uv(1-\alpha-\gamma)$	$\gamma v + v^2(1-\alpha-\gamma)$
with positive assortment	A	$\alpha u + u^2(1-\alpha)$	$uv(1-\alpha)$
	B	$uv(1-\gamma)$	$\gamma v + v^2(1-\gamma)$
with negative assortment	A	$u^2(1-\gamma)$	$\gamma u + uv(1-\gamma)$
	B	$\alpha v + uv(1-\alpha)$	$v^2(1-\alpha)$

		Frequencies of phenotypes of males mated monogamously	
		A	B
Without assortment	A	u^2	uv
	B	uv	v^2
With positive assortment	A	$\alpha u + \dfrac{u^2(1-\alpha)^2}{(1-\alpha u - \gamma v)}$	$\dfrac{uv(1-\alpha)(1-\gamma)}{(1-\alpha u - \gamma v)}$
	B	$\dfrac{uv(1-\alpha)(1-\gamma)}{(1-\alpha u - \gamma v)}$	$\gamma v + \dfrac{v^2(1-\gamma)^2}{(1-\alpha u - \gamma v)}$
With negative assortment	A	$\dfrac{u(u-\alpha v)(1-\gamma)}{(1-\alpha v - \gamma u)}$	$\gamma u + \dfrac{u(v-\gamma u)(1-\gamma)}{(1-\alpha v - \gamma u)}$
	B	$\alpha v + \dfrac{v(u-\alpha v)(1-\alpha)}{(1-\alpha v - \gamma u)}$	$\dfrac{v(v-\gamma u)(1-\alpha)}{(1-\alpha v - \gamma u)}$

When monogamous mating takes place without assortment of preferences the frequencies become those of random mating and no selection takes place. In assortative mating, the expression of female preference depends on whether the female herself possesses the preferred phenotype or not. With positive assortment, a preference is exercised only if the female has the same phenotype as the male she experimentally (see Ehrman, 1972). Hence they may be fitted to models in which frequencies are given by the expressions for $P_T(A)$ and $P_T(B)$. This model with complete expression of preference, which I call the 'C' model, has thus been fitted to data on frequencies of matings of different karyotypes of *Drosophila* (O'Donald, 1977a). Details of the procedure of fitting the 'C' model and other models will be given in the chapters in which the different models are analysed. Generally, as we shall see, all the models give rise to some frequency-dependence in the mating advantage gained by the males. In the 'C' models, the males take part in the same number of preferential matings when they are rare as when they are common; individually therefore, they mate more often when they are rare: the frequency-dependence is strongly negative. Polymorphisms can then be established at stable equilibrium frequencies.

3.2 Partial mating preferences

Various models can be set up with partial expression of the female mating preferences. A female's choice of mate must often be determined by the males she actually encounters before she is ready to mate. The chance of encounter with a particular phenotype depends on its frequency. The female who prefers the commoner phenotype will more often encounter a male of her choice than one who prefers a rare phenotype. The opportunity to express a preference will therefore depend on the frequency of the preferred males and the number of males that a female will encounter before she mates. Suppose that a female can choose her mate from a group of $n+1$ males with phenotypes A and B at relative frequencies x and y. Then the probability that such a group contains at

---

prefers: with negative assortment, the female must have a different phenotype to the male she prefers. When monogamous matings are assortative, frequencies of matings and genotypes differ from their frequencies produced by random mating, but there is still no selection of genes: gene frequencies remain constant. When matings are polygynous, the males who mated preferentially remain available to mate at random and selection of genes for the preferred characters takes place. Scudo and Karlin (1969) and Karlin (1969) analysed the effects of assortative mating with polygyny in models they called 'asymmetric models'. They also analysed the effects of assortative mating with monogamy in models in which the preferential matings either preceded or followed the random matings.

least one male of type A is $1-y^{n+1}$. Females with a preference for these males can therefore mate preferentially with this probability. If a proportion α of the females have this preference, the preferential matings with A males form a proportion $\alpha(1-y^{n+1})$ of all matings. The remaining proportion, αy^{n+1}, of these females do not meet a male they prefer and are therefore forced to mate with a B male. Similarly, if a proportion γ of the females prefer the B males, then $\gamma(1-x^{n+1})$ mate preferentially. The total frequencies of matings with A and B males are therefore given by the following equations (O'Donald, 1978a)

$$\left.\begin{array}{l} P_T(A)=\alpha(1-y^{n+1})+\gamma x^{n+1}+x(1-\alpha-\gamma) \\ P_T(B)=\alpha y^{n+1}+\gamma(1-x^{n+1})+y(1-\alpha-\gamma) \end{array}\right\} \qquad (3.2.1)$$

As $n \to \infty$, so the whole population of males becomes available for females to choose their mates from. The frequencies of matings are then the same as those of the 'C' model with complete preferences. The more general expressions for any value of n give frequencies of matings according to what I call the 'P' model with partial preferences. Thus in the 'C' model, the females with preferences go on searching indefinitely for a male with the phenotype they prefer: in the 'P' model, the search is abandoned after n males in the group have been encountered without success; mating takes place with male $n+1$ regardless of his phenotype.

The model with partial preferences can be derived in another way. This is based on the model described in section 2.4 of how females respond to male courtship. If some females have different thresholds of response towards different male phenotypes, they will mate preferentially by responding more quickly to some males than to others. Successive encounters with courting males raise a female's level of stimulation until a threshold has been reached at which she responds and mates at the next encounter. But if she has a higher threshold against particular male phenotypes, she will need extra encounters to reach this threshold. Having been stimulated to this higher level, she will then mate with the next male encountered regardless of his phenotype. If n is the number of encounters necessary to bridge the gap between the higher and lower thresholds, then a female with a lower threshold towards A males would mate with a B male only if she had had $n+1$ subsequent encounters with B males after having already been stimulated to the level sufficient for her to respond to an A male.

The probability that she makes these $n+1$ encounters with the 'wrong' kinds of male is y^{n+1}. She therefore mates preferentially with probability $1 - y^{n+1}$. If a proportion α of the females have this lower threshold towards the A males, then the proportion of preferential matings with A males is $\alpha(1 - y^{n+1})$. Thus we get the same expressions for $P_T(A)$ and $P_T(B)$ as when females choose their mates from a group of $n+1$ males. As $n \to \infty$ in this interpretation of the 'P' model, the gap between the thresholds becomes impossible to bridge: the preferred males possess some characteristic which is necessary to elicit the mating response of the females that prefer them thus giving rise to the 'C' models in the same way as before.

The expression of preference in the 'P' models is clearly frequency-dependent: the proportion of females expressing a preference for A males is $\alpha(1 - y^{n+1})$ which increases with the frequency, $1 - y$, of the preferred males. However, the mating advantage gained by the males is not so strongly dependent on the males' frequency as in the 'C' models. In the 'C' models, the proportion of preferential matings remains constant. In the 'P' models, however, rare males take part in a smaller proportion of preferential matings and so gain less of a frequency-dependent advantage. In models in which $n = 1$, rare males lose the whole of their frequency-dependent advantage. In fact, while rare they suffer a slight disadvantage and their selective advantage increases slightly with frequency.

This model can be generalized by allowing n to vary between females who have different preferences, for there is no reason to suppose that females with a preference for A males have their threshold for preferential mating reduced by the same amount as the females with a preference for B males. According to a model with different thresholds, females with a preference for A males mate at random after n encounters with B males: females with a preference for B mate at random after m encounters with A. Then the total frequencies of matings would become

$$P_T(A) = \alpha(1 - y^{n+1}) + \gamma x^{m+1} + x(1 - \alpha - \gamma)$$

$$P_T(B) = \alpha y^{n+1} + \gamma(1 - x^{m+1}) + y(1 - \alpha - \gamma)$$

In most of the models to be considered, however, we analyse only those in which $n = m$; but in Chapter 5, sections 5.1.1 and 5.2.1, some particular cases when $n \neq m$ are also analysed. The models with $n = m$ will apply generally when females exercise their choice from among a group of $n+1$ males since the size of the group will be the same for all females.

3.3 Frequency-dependent preferences

The 'P' models can be modified by assuming that preferences are expressed with a probability proportional to the frequency with which females encounter the males they prefer. In a group of $n+1$ males, this probability would vary from 0 for groups containing none of the preferred males up to 1 for groups consisting only of the preferred males. In this model, therefore, an increasing number of encounters with preferred males is assumed to facilitate the expression of mating preference. The preference exercised is both partial and positively frequency-dependent. Thus I call these models the 'PF +' models. In section 3.4, the corresponding models with negative frequency-dependence, or 'PF −' models are described.

In deriving the 'PF' models, I assume that females either make their choice from a group of $n+1$ males, or encounter males successively eventually mating after the $(n+1)$th encounter. They express their preferences according to either the number of preferred males in the group or the number encountered before mating. If, as before, the frequencies of A and B are x and y the probability that a group of $n+1$ males contains rA males is given by the binomial probability distribution

$$P_r = \binom{n+1}{r} x^r y^{n-r+1}$$

The probability that a female with a preference for A males expresses her preference and mates preferentially with an A male in this group is therefore

$$P'_r = \left(\frac{r}{n+1} \right) P_r$$

For all groups the proportion of females who mate preferentially is therefore

$$P'_m = \sum_{r=0}^{n+1} \left(\frac{r}{n+1} \right) P_r = x$$

The probability that these females do not express their preference but mate at random among the males is

$$P_m = \sum_{r=0}^{n+1} \left(\frac{n-r+1}{n+1} \right) P_r$$

and therefore they mate randomly with either A or B males with probabilities

$$P_m(A) = \sum_{r=0}^{n+1} \left[\frac{r(n-r+1)}{(n+1)^2} \right] P_r = nxy/(n+1)$$

and

$$P_m(B) = \sum_{r=0}^{n+1} \left[\frac{(n-r+1)^2}{(n+1)^2} \right] P_r = y(ny+1)/(n+1)$$

Among the females who prefer the A males a proportion $x+nxy/(n+1)$ therefore mate with A males and $y(ny+1)/(n+1)$ mate with B males. Since a proportion α of the females prefer A males, the total frequencies of matings with A and B males are given by the equations

$$P_T(A) = \alpha[x+nxy/(n+1)] + x(1-\alpha)$$
$$= x+\alpha nxy/(n+1)$$

and

$$P_T(B) = y - \alpha nxy/(n+1)$$

If a further proportion γ of the females prefer the B males, then we have total frequencies of matings

$$\left. \begin{array}{l} P_T(A) = x + nxy(\alpha-\gamma)/(n+1) \\ P_T(B) = y - nxy(\alpha-\gamma)/(n+1) \end{array} \right\} \tag{3.3.1}$$

As we should expect, when $n=0$ and females only have a choice of a single male, the matings are strictly proportional to the frequencies of the males. If equal proportions of females prefer each type of male ($\alpha=\gamma$), the matings are also proportional to frequencies. As $n\to\infty$, so the females then choose their mates from among the whole population of males and

$$\left. \begin{array}{l} P_T(A) = x + xy(\alpha-\gamma) \\ P_T(B) = y - xy(\alpha-\gamma) \end{array} \right\} \tag{3.3.2}$$

These frequencies can also be derived from those of the 'C' model with complete preferences in which we have (equations 3.1.1)

$$P_T(A) = \alpha + x(1-\alpha-\gamma)$$
$$P_T(B) = \gamma + y(1-\alpha-\gamma)$$

If α and γ are assumed to be proportional to frequency, we can substitute αx and γy in place of α and γ in the equations for $P_T(A)$ and $P_T(B)$ and obtain

$$P_T(A) = \alpha x + x(1 - \alpha x - \gamma y)$$
$$= x + xy(\alpha - \gamma)$$
$$P_T(B) = y - xy(\alpha - \gamma)$$

which are equations 3.3.2 for the 'PF+' model $(n \to \infty)$. The same frequencies can also be derived from the 'P' model with partial preferences by putting $n = 1$. Females are then choosing their mates from between two males and mating preferentially if at least one of the males has the preferred phenotype. In general in the 'PF+' models, the mating advantage gained by the males increases as they increase in frequency: the positive frequency-dependence of the preferences produces a positive frequency-dependence in the selective advantage. Thus, as we shall show in the analysis of these models, stable polymorphisms can never be maintained. This is also true of the 'P' models when $n = 1$.

3.4   Negative frequency-dependence

Spiess and Ehrman (1978) suggested that females may become habituated towards the more common males and then only respond to other rarer males. They used this suggestion to explain the frequency-dependent mating advantage of rare male *Drosophila* which had been demonstrated in many of their own experiments (Ehrman, 1967, 1968, 1969, 1972; Spiess, 1968; Ehrman and Spiess, 1969; Spiess and Spiess, 1969; Spiess and Ehrman 1978). The data can just as well be fitted to the 'C' models (O'Donald, 1977a), since the constant proportion of preferential matings in these models entails strong frequency-dependence in mating success. As we shall show in Chapter 5, the 'P' models fit some of the data even better. However, Spiess and Ehrman's suggestion can be formulated quantitatively in terms of preferences for particular phenotypes, giving rise to models with negative frequency-dependence in the mating preferences – the 'PF−' models. These models too can be fitted to the data.

Suppose a female may become habituated towards B males with probability α; the probability that she does become habituated depends on the number of B males she encounters. The probability of habituation takes the values $0, 1/(n+1), \ldots, n/(n+1)$, and 1 as the number of B males in a group of $n+1$ males varies from 0 to $n+1$. These probabilities of becoming habituated are the probabilities that she will not mate with a B male in the group of males from which she will choose her mate.

The parameter α may equally well represent a proportion of females who prefer to mate with A males and whose expression of this preference decreases with increasing numbers of A males in the group. According to this idea, the females who prefer A males (or who may become habituated towards B and hence inhibited from mating with B) mate preferentially with a probability

$$P'_m = \sum_{r=0}^{n+1} \left(\frac{n-r+1}{n+1}\right) P_r = y$$

They mate randomly with A and B males with probabilities

$$P_m(A) = \sum_{r=0}^{n+1} \left[\frac{r^2}{(n+1)^2}\right] P_r = x(nx+1)/(n+1)$$

$$P_m(B) = \sum_{r=0}^{n+1} \left[\frac{r(n-r+1)}{(n+1)^2}\right] P_r = nxy/(n+1)$$

Therefore for all matings, the frequencies of matings with A and B males are

$$P_T(A) = \alpha[y + x(nx+1)/(n+1)] + x(1-\alpha)$$
$$= x + \alpha y(ny+1)/(n+1)$$
$$P_T(B) = y - \alpha y(ny+1)/(n+1)$$

If we include a proportion γ of females who prefer B males and whose preference for them is expressed with negative frequency-dependence (or, alternatively, females who become habituated to A males with probability γ) then the total frequencies of matings with A and B males are given by the equations

$$\left. \begin{aligned} P_T(A) &= x + [\alpha y(ny+1) - \gamma x(nx+1)]/(n+1) \\ P_T(B) &= y - [\alpha y(ny+1) - \gamma x(nx+1)]/(n+1) \end{aligned} \right\} \qquad (3.4.1)$$

As $n \to \infty$, these frequencies become

$$\left. \begin{aligned} P_T(A) &= x + \alpha y^2 - \gamma x^2 \\ P_T(B) &= y - \alpha y^2 + \gamma x^2 \end{aligned} \right\} \qquad (3.4.2)$$

They can also be derived from the corresponding frequencies in the 'C' model by substituting αy in place of α and γx in place of γ in equations 3.1.1, thus putting the expression of preference proportional to the frequencies of those phenotypes that are not the objects of the preferences. In this model, the males' mating success increases as they become less common: both the mating success and the mating preference are thus negatively frequency-dependent. When $\alpha = \gamma$, the frequencies of matings in the 'PF−' model become the same as those in the 'C' model, giving rise to the same level of frequency-dependence in spite of the completely different assumptions that are made in the derivation of the two models. If preferences for two phenotypes are roughly equal, data on numbers of matings are not likely to distinguish between the 'PF−' and 'C' models: both would give a roughly similar fit to data of numbers of matings.

A physiological mechanism may determine mating preferences that act with negative frequency-dependence. In Averhoff and Richardson's model (discussed in Chapter 2, section 2.4), the female response in *Drosophila melanogaster* depends on two factors: a non-volatile factor, transmitted to females in close association with courting males, and a highly volatile pheromone. Averhoff and Richardson stated that the 'existence of both long-range volatile pheromones and short-range slightly volatile or nonvolatile signals could explain some of the nonlinear details of density and proportions which Ehrman and Spiess have reported in the minority male advantage'. This obscure statement gives no suggestion of any specific mechanism for the frequency-dependent expression of mating preference. Let us suppose, however, that a proportion α of the females require a non-volatile signal from a particular genotype before they can respond to the volatile pheromone. According to Averhoff and Richardson, the concentration of the volatile pheromone reaching the receptor cells then determines the females' responsiveness. The females may become less responsive at higher concentrations if the receptors accommodate to substances that constantly bombard them: higher concentrations would thus produce inhibition. If this inhibition acts with a typical dosage–response relationship, such as may be fitted by probit analysis, then larger numbers of males with a particular genotype producing higher concentrations of their own volatile pheromone will inhibit an increased proportion of the females. If the pheromone concentrations lie within a range of values in which an approximately linear increase takes place in the proportion of females inhibited, then the frequencies of matings will be given by those of the 'PF $-$' model.

If there was no upper threshold at which inhibition occurred, but only a lower threshold required for response, then this mechanism would give rise to mating frequencies according to the 'PF $+$' model.

3.5   Competition between males

Males usually compete between themselves for mates: they fight, threaten, defend territories and try to outdo one another in sexual display. Sexual selection will take place if males vary in their abilities to do these things, which are often closely correlated. As we have shown (Chapter 2, section 2.3), in birds threatening behaviour towards the other males and sexual display towards the females are determined by levels of testosterone, which must directly determine the chances of mating. For example, suppose the sexually more active males produce k times the stimulus of the less active males: a female who encounters one of the more active males is stimulated by an amount x and she then mates with one of these males if she has already been stimulated to within x of her threshold of response. The other males only stimulate the females by an amount x/k. Among the females who are within x of their threshold, a proportion $(x/k)/x = 1/k$ will also be within the point x/k of their threshold and will mate on encountering one of the less active males. The remaining females, who represent a proportion $(k-1)/k$, will only mate with the less active males after one or more extra encounters with them. We may put $(k-1)/k = \alpha$, for this is the proportion of females who act as if they had a mating preference for the more active males. If $k < 2$, only one extra encounter will be needed before all remaining females mate at random with any male. This model of male competition in sexual display would thus be formally equivalent to the 'P' model of partial preference with $n = 1$. If $k > 2$, some females would require more than one extra encounter giving rise to a mixed 'P' model with $n = 1$ for some females and $n = 2, 3, \ldots$ for others. Since it is unlikely that males would differ genetically by a very large amount in their level of sexual activity, the 'P' model with $n = 1$ should adequately represent this form of male competition in most cases: some males might then differ genetically from others by a factor of up to 2 in their level of sexual activity.

A well as displaying more actively towards the females, some males may also behave more aggressively towards other males, preventing them from setting up territories and finding mates. As suggested in Chapter 2, section 2.3, and demonstrated in Chapter 7, section 7.5, a male who can defend a larger part of the breeding grounds as his territory thus reduces the size of other males' territories and increases his own chances of mating. If the females tend to arrive successively and at random on the breeding grounds, they will be more likely to land on the larger territories than on the smaller. The probability of mating will then be proportional to the ratio of the territory sizes. In section 7.5, I show that in a monogamous species of bird the males with the larger territories will mate at an earlier average date in the breeding season. This earlier mating will give them a selective advantage because the earlier breeders among the pairs of birds will have greater success in fledging their young. In a polygynous species, the males with the larger territories will mate more often than the others with the smaller territories. Charlesworth and Charlesworth (1975) described a model of sexual selection which would apply generally to this sort of competition between males. They postulate that the male with the sexually advantageous phenotype mates k times more often than the other males. Thus in a group of n males of which n_A are of the sexually advantageous type A, the probability that an A male succeeds in mating is

$$kn_A/(n-n_A+kn_A).$$

Therefore, if p is the frequency of A males, then the average number of matings an A male achieves in competition in groups of n males is

$$\left(\frac{1}{p}\right)\sum_{n_A=1}^{n}\left\{\left[\binom{n}{n_A}p^{n_A}(1-p)^{n-n_A}\right]\left[kn_A/(n-n_A+kn_A)\right]\right\} \quad (3.5.1)$$

Charlesworth and Charlesworth call this quantity the mean fertility of the A males. The mean fertility of the other males will then be determined in a similar way by the probability, $(n-n_A)/(n-n_A+kn_A)$, that one of the other males succeeds in mating. Unfortunately the mean fertilities cannot be formulated in simple mathematical terms, but given values of k, n and p the ratios of the fertilities can be calculated by computer. A computer can therefore be used to determine the outcome of selection according to this model of male competition. The model produces a slight positive frequency-dependence in the fertility ratios. Both *A* and *B* have the highest fertilities when *A* is rare and *B* common. The fertilities decline as *A* increases in frequency; but the ratio of the fertilities remains more or less constant, *A* gaining only a slight increase of advantage with frequency. If *A* is very rare and groups of *n* males contain either one *A* male or none, then Charlesworth and Charlesworth show that as $p \to 0$, the ratio of the fertility of *B* to that of *A* becomes

$$(1/n) \ [(n-1)/k+1]$$

which is always less than 1 if *k* is greater than 1. Type *A* is initially advantageous and always remains so: *A*'s selective advantage increases slightly as *A* increases in frequency. As a result of this positive frequency-dependence, no polymorphisms can be maintained unless the sexually superior male is heterozygous. Otherwise as shown in section 3.6 which follows, sexual selection must take the *A* males to complete fixation.

3.6   Example of genetical analysis of a model of sexual selection

In the Charlesworths' model described in the previous section, males of one phenotype are *k* times more likely to find mates than males of another phenotype. In a group of males competing for mates, n_A are phenotypically *A* and $n - n_A$ are phenotypically *B*. If the *A* males are the better competitors, then they mate with a probability given by expression 3.5.1. A similar expression gives the *B* males' probability of mating. The values of these probabilities of mating can then be computed for particular values of *n*. If a large number of males compete, however, the phenotypic frequencies in the groups will be the same as their frequencies in the population. Simple analytical expressions can then be found for the changes in gene frequency according to particular models of the genetic determination of the phenotypes. A genetical analysis of this simple model of male competition will exemplify the general method I have used to analyse the models described in Chapters 4, 5 and 6.

If the phenotypes *A* and *B* are determined by two alleles at a locus, then *A* may be dominant or recessive or heterozygous. For generality, if *A* is heterozygous, a second parameter must be introduced for the competitive ability of one homozygote relative

Table 3.2 *Probabilities of matings of males according to different genetical models*

Genetic model	Probabilities of mating of males of given genotype		
	$P_T(AA)$	$P_T(Aa)$	$P_T(aa)$
A dominant	$\dfrac{ku}{k-w(k-1)}$	$\dfrac{kv}{k-w(k-1)}$	$\dfrac{w}{k-w(k-1)}$
A recessive	$\dfrac{ku}{1+u(k-1)}$	$\dfrac{v}{1+u(k-1)}$	$\dfrac{w}{1+u(k-1)}$
No dominance	$\dfrac{ku}{ku+lv+w}$	$\dfrac{lv}{ku+lv+w}$	$\dfrac{w}{ku+lv+w}$

In this table the genotypes AA, Aa and aa are present in the population at frequencies u, v and w. $P_T(AA)$, $P_T(Aa)$ and $P_T(aa)$ are the probabilities that males with these genotypes will find mates.

to the other: parameters k and l can be defined as the number of times that the genotypes AA and Aa mate relative to the genotype aa subject to the condition $k,l > 1$. If $k = 1$, the two homozygotes are equally inferior to the heterozygotes in their competitive ability. Table 3.2 shows the probabilities of the matings of the males according to the three genetical models. These probabilities then give the frequencies of the matings between different genotypes shown in Table 3.3. In the generation in which the matings take place, the genotypes AA, Aa and aa occur at frequencies u, v and w in the population. In the next generation, their frequencies are transformed by sexual selection and genetic segregation into the frequencies u', v' and w'.

A is dominant.   When A is dominant, the matings shown in table 3.3 give rise to genotypes with frequencies in the next generation

$$u' = kp^2/[k-w(k-1)]$$

$$v' = p(kv+kw+w)/[k-w(k-1)]$$

$$w' = q(w+\tfrac{1}{2}kv)/[k-w(k-1)]$$

In these equations p and q are the gene frequencies of the alleles A and a defined by $p = u+\tfrac{1}{2}v$ and $q = w+\tfrac{1}{2}v$. In the next generation the allele A occurs at frequency

$$p' = p[k-\tfrac{1}{2}w(k-1)]/[k-w(k-1)]$$

Hence the change in gene frequency from one generation to the

Table 3.3   *Frequencies of matings between genotypes after competition between males*

	Frequencies of matings		
Matings	A dominant	A recessive	No dominance
$AA \times AA$	$ku^2/[k-w(k-1)]$	$ku^2/[1+u(k-1)]$	$ku^2/(ku+lv+w)$
$AA \times Aa$	$2kuv/[k-w(k-1)]$	$(k+1)uv/[1+u(k-1)]$	$(k+l)uv/(ku+lv+w)$
$AA \times aa$	$(k+1)uw/[k-w(k-1)]$	$(k+1)uw/[1+u(k-1)]$	$(k+1)uw/(ku+lv+w)$
$Aa \times Aa$	$kv^2/[k-w(k-1)]$	$v^2/[1+u(k-1)]$	$lv^2/(ku+lv+w)$
$Aa \times aa$	$(k+1)vw/[k-w(k-1)]$	$2vw/[1+u(k-1)]$	$(l+1)vw/(ku+lv+w)$
$aa \times aa$	$w^2/[k-w(k-1)]$	$w^2/[1+u(k-1)]$	$w^2/(ku+lv+w)$

next is given by

$$\Delta p = \tfrac{1}{2}pw(k-1)/[k-w(k-1)]$$

This shows that $\Delta p > 0$ for all $k > 1$. Therefore in the course of evolution $p \to 1$ and the genotype AA is ultimately fixed in the population. There is no frequency-dependence in the advantage gained by the A males. The relative frequencies of matings of males phenotypically A compared with males phenotypically B (genotypically aa) are simply $k/[k-w(k-1)]$ and $1/[k-w(k-1)]$, and hence always in the ratio $k:1$.

A is recessive.   The genotypic frequencies in the next generation are given by the equations

$$u' = p(ku + \tfrac{1}{2}v)/[1+u(k-1)]$$

$$v' = q(ku + u + v)/[1+u(k-1)]$$

$$w' = q^2/[1+u(k-1)]$$

We then find the change in gene frequency from one generation to the next

$$\Delta p = \tfrac{1}{2}qu(k-1)/[1+u(k-1)]$$

As when A is dominant, $\Delta p > 0$ for all $k > 1$ and hence $p \to 1$ in the course of evolution. The constant advantage of one phenotype over the other ensures fixation of the advantageous type.

A shows no dominance.   Heterozygotes compete separately with homozygotes in the search for mates. Table 3.3 shows the frequencies of the matings. Genotypes are then produced in the next generation with frequencies

$$
\left.\begin{array}{l}
u' = p(ku + \tfrac{1}{2}lv)/(ku + lv + w) \\[4pt]
v' = (kuq + \tfrac{1}{2}lv + wp)/(ku + lv + w) \\[4pt]
w' = q(w + \tfrac{1}{2}lv)/(ku + lv + w)
\end{array}\right\} \tag{3.6.1}
$$

The change in gene frequency is now

$$
\Delta p = \tfrac{1}{2}[q(ku + \tfrac{1}{2}lv) - p(w + \tfrac{1}{2}lv)]/(ku + lv + w) \tag{3.6.2}
$$

showing that at equilibrium when $\Delta p = 0$ we have the relation

$$
q(ku + \tfrac{1}{2}lv) = p(w + \tfrac{1}{2}lv)
$$

and hence

$$
p = (ku + \tfrac{1}{2}lv)/(ku + lv + w) \tag{3.6.3}
$$

Putting $u' = u$, $v' = v$ and $w' = w$ and substituting expression 3.6.3 for p in equations 3.6.1, we find that at equilibrium $u = p^2$, $v = 2pq$ and $w = q^2$. After equilibrium has been reached, the genotypes are therefore maintained at the Hardy–Weinberg frequencies. Thus substituting these frequencies for u, v and w in expression 3.6.3 we get

$$
p = (kp^2 + lpq)/(kp^2 + 2lpq + q^2)
$$

This equation can be factorized to give

$$
q[(k - 2l + 1)p + l - 1] = 0
$$

showing that there will be an equilibrium with the allele A at a frequency

$$
p_* = (l - 1)/(2l - k - 1)
$$

A polymorphism will therefore exist if $1 > p_* > 0$ or if $lk > 1$.

The local stability of such a polymorphism can be tested by the following method, which has been used throughout this book in testing the stability of polymorphisms maintained by sexual selection. In all models of sexual selection in which there is no assortative mating (for example, the models described in Chapters 4, 5 and 6), the genotypes are maintained at Hardy–Weinberg frequencies after equilibrium has been reached. In the region of the equilibrium point, genotypic frequencies approximate closely to those of Hardy–Weinberg. To test for local stability of an equilibrium, we can therefore assume Hardy–Weinberg frequencies exist and substitute these in the expression for Δp. Making these substitutions in equation 3.6.2 we get the equation

$$\Delta p = \tfrac{1}{2}pq[l-1-p(2l-k-1)]/(ku+lv+w)$$
$$= -\tfrac{1}{2}(2l-k-1)pq(p-p_*)/(ku+lv+w)$$

The equilibrium is then stable, if the rate of change of Δp is negative in the region of the equilibrium point: if so, a deviation from the equilibrium will be progressively reduced in subsequent generations ultimately restoring the population to equilibrium. Differentiating Δp, we get

$$\frac{d\Delta p}{dp} = -\tfrac{1}{2}(2l-k-1)\left[\frac{pq}{ku+lv+w}+(p-p_*)\frac{d}{dp}\left(\frac{pq}{ku+lv+w}\right)\right]$$

Hence at $p=p_*$,

$$\frac{d\Delta p}{dp} = -\tfrac{1}{2}(2l-k-1)p_*q_*/(ku_*+lv_*+w_*)$$

The equilibrium is a stable point if $2l-k-1>0$ or $l>\tfrac{1}{2}(k+1)$. The condition for existence of the polymorphism therefore implies the condition for its stability. The heterozygote must be a better competitor for mates than either of the homozygotes: heterozygote advantage must prevail. This confirms the conclusion reached in section 3.5 following discussion of D. and B. Charlesworth's model.

The Charlesworths' model is a general one, implying no particular mechanism by which the males compete and increase their chances of finding a mate. Male competition can be a complex interaction, however, taking place in several stages. This has been demonstrated in Parker's model of sexual selection in the dung fly, *Scatophaga stercoraria* (Parker, 1970, 1974a,b; and see Chapter 2, section 2.2). The males search for the females on the surface of fresh droppings of dung. Some males, having found a female are dislodged by others, either during copulation or oviposition. Since the last male to mate with a female fertilizes 81.4 per cent of her next batch of eggs on average, he gains a considerable advantage. The percentage fertilized also varies with the duration of copulation. As previously explained, the advantage gained is a complicated function of the duration of copulation and the probabilities of mating with virgin, non-virgin, copulating or ovipositing females; it is not just a simple function of the relative frequency of mating as in the Charlesworths' model.

We may expect that selective values will often be complex functions of competitive behaviour, unique to the particular species. Genetic variation in some aspect of the behaviour will then give rise to some specific selective advantage or disadvantage. A model of the selection will apply only to the species in which the behaviour has been studied. Parker uses his own model of sexual selection in the dung fly to show that the actual behaviour is fairly close to the optimum that would maximize fitness. He then argues that the males' competitive behaviour has evolved by sexual selection for this optimum. He does not analyse a genetical model, however, and his argument has the weakness that selection does not necessarily maximize fitness unless reproduction is asexual or gene action is additive. If genetic interaction produces linkage disequilibrium, maximum fitness may not even be attainable: mean fitness often declines from a maximum as a result of interactive selection. If different aspects of behaviour are determined by genes at different loci, then inevitably they will interact in their effect on fitness. The assumption that fitness will maximize, taking each aspect of behaviour to an optimum, is therefore unwarranted. As we shall see in a genetical model of the evolution of mating preferences (Chapter 8, section 8.4), the outcome of selection may depend entirely on the initial frequencies, since these determine the level of linkage disequilibrium that can arise and the domain of attraction to equilibrium.

I cast no doubt on the validity of Parker's analysis of the selective forces that are the product of the males' competitive behaviour, but rather a doubt whether these selective forces necessarily entail the evolution of optimum behaviour. When no specific genetic model has been formulated, there must always be doubt about predictions of the evolutionary consequences of complex or interactive selection: such predictions are then really no more than guesses. In my own approach to the study of sexual selection, I start with mechanisms of mating behaviour and mate selection that are simple and general rather than detailed and specific. These general mechanisms are the models that I have described in this chapter. Following the example given in this section, I then deduce their evolutionary consequences by making explicit assumptions about the genetic determination of the characteristics that the selection acts upon.

4

Complete expression of mating preference

4.1 Preferences for dominant and recessive characters

In describing in Chapter 3 how mating preferences among the females determine the frequencies of the matings of the males, I assumed that two phenotypes of males, A and B, were preferred by proportions α and γ of the females. I now analyse the genetical consequences of mating preferences that are always expressed whenever females can choose between different male phenotypes in the population: these are the models with complete preferences, or 'C' models, of Chapter 3. The frequencies of matings of males with A and B phenotypes are then given by

$$P_T(A) = \alpha + x(1 - \alpha - \gamma)$$
$$P_T(B) = \gamma + y(1 - \alpha - \gamma)$$

where x and y are the frequencies of A and B. I shall assume in this section that A is a dominant, determined by the genotypes AA or Aa, and B is the recessive aa. If u, v and w are the frequencies of AA, Aa and aa, then

$$x = u + v = 1 - w$$
$$y = w$$

and

$$\left.\begin{aligned}
P_T(AA) &= \alpha u/(1 - w) + u(1 - \alpha - \gamma)\\
P_T(Aa) &= \alpha v/(1 - w) + v(1 - \alpha - \gamma)\\
P_T(aa) &= \gamma + w(1 - \alpha - \gamma)
\end{aligned}\right\} \qquad (4.1.1.)$$

These are the frequencies with which the females mate with each genotype. Hence the frequencies of the matings are shown in table 4.1. In the next generation, the genotypic frequencies are given by

Table 4.1 *Frequencies of matings as a result of sexual selection with complete preferences*

Mating type	Frequency of mating type	
	Preference for dominant and recessive characters	Separate preference for each genotype
$AA \times AA$	$\alpha u^2/(1-w)+(1-\alpha-\gamma)u^2$	$\alpha u+(1-\alpha-\beta-\gamma)u^2$
$AA \times Aa$	$2\alpha uv/(1-w)+2(1-\alpha-\gamma)uv$	$\alpha v+\beta u+2(1-\alpha-\beta-\gamma)uv$
$AA \times aa$	$\alpha uw/(1-w)+\gamma u+2(1-\alpha-\gamma)uw$	$\alpha w+\gamma u+2(1-\alpha-\beta-\gamma)uw$
$Aa \times Aa$	$\alpha v^2/(1-w)+(1-\alpha-\gamma)v^2$	$\beta v+(1-\alpha-\beta-\gamma)v^2$
$Aa \times aa$	$\alpha vw/(1-w)+\gamma v+2(1-\alpha-\gamma)vw$	$\beta w+\gamma v+2(1-\alpha-\beta-\gamma)vw$
$aa \times aa$	$\gamma w+(1-\alpha-\gamma)w^2$	$\gamma w+(1-\alpha-\beta-\gamma)w^2$

With separate preference for each genotype α, β and γ are the proportions of females who prefer males with genotypes AA, Aa and aa. With dominance α is the proportion who prefer AA and Aa males indiscriminately and hence mate at random among the males with these genotypes; γ is the proportion who prefer aa males. The genotypes AA, Aa and aa have frequencies u, v and w in both males and females.

the recurrence equations

$$\left.\begin{aligned}
u' &= \alpha p^2/(1-w)+p^2(1-\alpha-\gamma) \\
v' &= \alpha p(2q-w)/(1-w)+\gamma p+2pq(1-\alpha-\gamma) \\
w' &= \alpha q(q-w)/(1-w)+\gamma q+q^2(1-\alpha-\gamma)
\end{aligned}\right\} \quad (4.1.2)$$

where p and q are the frequencies of the alleles A and a. Therefore in the next generation

$$p' = p+\tfrac{1}{2}p[-\gamma+w(\alpha+\gamma)]/(1-w)$$

or expressed as the finite difference of the change from one generation to the next

$$\Delta p = \tfrac{1}{2}p[-\gamma+w(\alpha+\gamma)]/(1-w) \quad (4.1.3)$$

There is an equilibrium point at a frequency of the recessive phenotype given by

$$w_* = \gamma/(\alpha+\gamma) \quad (4.1.4)$$

Since

$$w' = q^2-pq[-\gamma+w(\alpha-\gamma)]/(1-w)$$

therefore, at equilibrium, when

$$w' = w = w_*$$

$$w_* = q_*^2$$

Similarly

$$v_* = 2p_*q_*$$

and

$$u_* = p_*^2$$

showing that at the equilibrium the genotypes are found to occur in the Hardy–Weinberg frequencies. Thus

$$q_* = \sqrt{[\gamma/(\alpha+\gamma)]}$$
$$p_* = 1 - \sqrt{[\gamma/(\alpha+\gamma)]} \qquad (4.1.5)$$

Since we may write equation 4.1.3 in the form

$$\Delta p = \tfrac{1}{2}p(\alpha+\gamma)(w-w_*)/(1-w)$$

and assume near equilibrium that

$$w = q^2, \qquad \frac{dw}{dp} = -2q$$

therefore

$$d\Delta p/dp|_{p=p_*} = -p_*q_*(\alpha+\gamma)/(1-w_*)$$

This is always negative for positive values of α and γ and hence the polymorphic equilibrium state is stable.

For slow changes in gene frequency, the finite difference

$$\Delta p = \tfrac{1}{2}p(\alpha+\gamma)(q^2-q_*^2)/(1-q^2)$$

is a good approximation to a differential equation. It has the general solution

$$\left(\frac{q_T-q_*}{q_0-q_*}\right)^{(q_*+1)} \cdot \left(\frac{q_T+q_*}{q_0+q_*}\right)^{(q_*-1)} = \exp[-Tq_*(\alpha+\gamma)] \qquad (4.1.6)$$

where q_0 is the frequency of the allele a at generation zero and q_T is its frequency T generations later.

When females have preferences for only one of the phenotypes, the corresponding allele evolves to fixation. When only the dominant is preferred ($\gamma=0$), $p_*=1$, and the general solution of the differential equation is not applicable. However,

$$\Delta p = \tfrac{1}{2}\alpha pw/(1-w)$$

and approximately

$$\Delta p = \tfrac{1}{2}\alpha q^2/(1+q)$$

As a differential equation, this has the solution

$$\log_e(q_0/q_T) + 1/q_T - 1/q_0 = \tfrac{1}{2}\alpha T$$

When only the recessive is preferred, then $q_* = 1$, and the general solution of the differential equation (given by 4.1.6.) reduces to

$$q_T = 1 - p_0 e^{-\frac{1}{2}\gamma T}$$

In this case an exact solution of the recurrence equation can be found, since

$$p' = p - \tfrac{1}{2}\gamma p$$
$$p_T = p_0(1 - \tfrac{1}{2}\gamma)^T$$
$$q_T = 1 - p_0(1 - \tfrac{1}{2}\gamma)^T$$

showing that the solution of the differential equation is only valid for small values of γ. Table 4.2 shows the percentage error in the solution of the differential equation at certain magnitudes of the total female preference. For a few generations the error remains small even when as many as 20 per cent of the females express preferences for the male phenotypes. Equation 4.1.6 for the change in gene frequency can therefore be used to estimate α and γ from the gene frequencies in successive generations in a population cage. Given trial values of q_* and $\alpha + \gamma$, values of

$$(q_* + 1)\log(q_T - q_*) + (q_* - 1)\log(q_T + q_*)$$

can be fitted by least squares to their hypothetical values given by

Table 4.2 *Percentage error in equation 4.1.6 when mating preferences produce a stable equilibrium at a frequency* $q_* = 0.5$

No. of generations	$\alpha + \gamma = 0.02$	Percentage error $\alpha + \gamma = 0.04$	$\alpha + \gamma = 0.20$
10	0.03	0.15	3.82
20	0.87	0.30	7.72
30	0.11	0.46	11.67
40	0.15	0.61	15.70
50	0.19	0.76	19.85
100	0.38	1.49	42.73

$- Tq_*(\alpha+\gamma)+C$ where C is a constant. In this way improved estimates of q_* and $(\alpha+\gamma)$ will be obtained for the next iterative step and so on until the best fit has been obtained.

It is interesting to compare the rates of sexual selection and natural selection in corresponding models. In sexual selection, if a dominant is the preferred phenotype, then we have the difference equation

$$\Delta p = \tfrac{1}{2}\alpha pw/(1-w)$$

for comparison with the corresponding difference equation for the natural selection of a dominant

$$\Delta p = spq^2/(1-sq^2)$$

where s is the selective coefficient of the relative disadvantage of the recessive phenotype. Compared with natural selection determined by a selective coefficient $s=\tfrac{1}{2}\alpha$, sexual selection is much faster when the allele A is rare but slows down to a similar rate to natural selection when A is common. Figure 4.1 shows how the rates of sexual and natural selection may differ (O'Donald, 1977c).

4.2 Separate preferences for each genotype

Table 4.1 shows the expected frequencies of matings when females have preferences for each genotype separately: α, β and γ are now their preferences for the males with genotypes AA, Aa and aa. The recurrence equations for the genotypic frequencies are as follows (O'Donald, 1973b).

$$\left.\begin{aligned} AA:\ u' &= p(\alpha+\tfrac{1}{2}\beta)+p^2(1-\alpha-\beta-\gamma)\\ Aa:\ v' &= p(\gamma+\tfrac{1}{2}\beta)+q(\alpha+\tfrac{1}{2}\beta)+2pq(1-\alpha-\beta-\gamma)\\ aa:\ w' &= q(\gamma+\tfrac{1}{2}\beta)+q^2(1-\alpha-\beta-\gamma) \end{aligned}\right\} \quad (4.2.1)$$

Thus the recurrence equation for the gene frequency of A is

$$p' = \tfrac{1}{2}(\alpha+\tfrac{1}{2}\beta)+p(1-\tfrac{1}{2}\alpha-\tfrac{1}{2}\beta-\tfrac{1}{2}\gamma)$$

If we put $\phi=\alpha+\tfrac{1}{2}\beta$ and $\theta=\alpha+\beta+\gamma$, then

$$p' = \tfrac{1}{2}\phi + p(1-\tfrac{1}{2}\theta) \quad (4.2.2)$$

$$p_* = \phi/\theta$$

The equilibrium frequency, p_*, is approached at a rate given by the

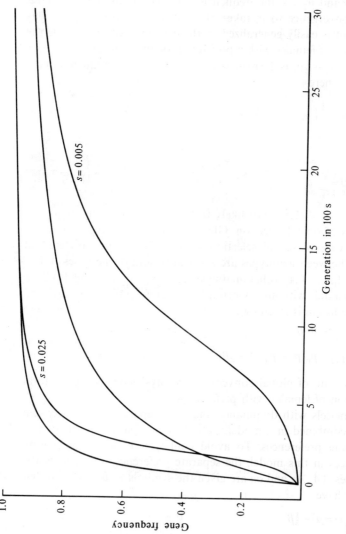

Figure 4.1 Gene frequencies of a dominant allele increasing in a population either by sexual selection or by natural selection. The natural selective coefficients were $s = 0.025$ and $s = 0.005$ and are shown alongside the graphs of the gene frequencies in successive generations. For both examples of natural selection, the corresponding sexual selection is determined by a mating preference $\alpha = 2s$ (see text). The sexual selection is initially much faster than the natural selection, but the final rates

general solution of the recurrence equation

$$p_T = p_* - (p_* - p_0)(1 - \tfrac{1}{2}\theta)^T \tag{4.2.3}$$

where p_0 and p_T are the frequencies in generations 0 and T. Thus global convergence to p_* takes place at a geometric rate $(1 - \tfrac{1}{2}\theta)$. This result is easily generalized to the case of n alleles. If α_{ij} is the proportion of females with a preference for the genotype $A_i A_j$, then $p_i' = \tfrac{1}{2}\phi_i + p_i(1 - \tfrac{1}{2}\theta)$ is the recurrence equation for the frequency of the ith allele, where

$$\phi_i = \frac{1}{2}\left(\alpha_{ii} + \sum_j \alpha_{ij}\right)$$

and

$$\theta = \sum_{i,j} \alpha_{ij}.$$

The use of n alleles is not likely to be relevant to many biological examples of sexual selection. Ghai (1974) analysed the more complicated case of sexual selection with n alleles when preferential matings between genotypes are assortative and of the types $A_i A_i \times A_i A_i$ or $A_i A_j \times A_i A_j$, other matings occurring at random. For the simple model without assortment of preferential matings, the general solution is given by

$$p_{i*} = \phi_i / \theta$$

$$p_{iT} = p_{i*}(p_{i*} - p_{i0})(1 - \tfrac{1}{2}\theta)^T$$

Thus the rate of global convergence always depends on the total proportion of females with preferences.

The models with dominance and with separate preferences can also be combined in a model that allows for partial dominance of the mating preferences. To avoid confusion between the mating preferences in this model, the separate preferences for each of the genotypes AA, Aa and aa are given the symbols α', β' and γ' so that we now have

$$\phi = \alpha' + \tfrac{1}{2}\beta'$$

$$\theta = \alpha' + \beta' + \gamma'$$

Then α is the preference for A as a dominant allele. Thus in the combined model the total proportion of females with preferences is

$\alpha + \theta$. In this way a model of partial dominance can be defined; for, if $\beta' = \gamma' = 0$, then $\phi = \theta = \alpha'$ which is the preference for AA males while α is the preference for AA or Aa. Thus some females are sufficiently stimulated to respond to both the AA and the Aa males: others are sufficiently stimulated only by AA males. The greater stimulation by the AA males provides an additional mating preference in their favour. In general, when α is the parameter of the preference for either AA or Aa and $\phi = \alpha' + \frac{1}{2}\beta'$, $\theta = \alpha' + \beta' + \gamma'$ are the parameters of the preferences for each genotype separately, then we have the recurrence equations

$$\left. \begin{aligned} u' &= \phi p + \alpha p^2/(1-w) + p^2(1-\alpha-\theta) \\ v' &= \phi q + (\theta - \phi)p + \alpha p(2q-w)/(1-w) + 2pq(1-\alpha-\theta) \\ w' &= (\theta - \phi)q + \alpha q(q-w)/(1-w) + q^2(1-\alpha-\theta) \end{aligned} \right\} (4.2.4)$$

At equilibrium putting $p' = p$ and $w' = w$

$$p_* = \phi(1-w)/(\theta - \alpha w - \theta w)$$

$$w(1-w) = pq(\theta - \alpha w - \theta w) - \phi q(1-w) + q^2(1-w)$$

so that $w = q^2$
giving the cubic equation

$$q^3(\alpha + \theta) - q^2(\alpha + \theta - \phi) - \theta q + \theta - \phi = 0$$

or

$$(q-1)[q^2(\alpha + \theta) + \phi q - \theta + \phi] = 0$$

A stable polymorphic equilibrium is then established at the following frequency of the allele a:

$$q_* = \frac{-\phi + \sqrt{[\phi^2 + 4(\theta - \phi)(\alpha + \theta)]}}{2(\alpha + \theta)} \qquad (4.2.5)$$

When $\alpha' = \beta' = 0$, so that $\phi = 0$ and $\theta = \gamma'$, then the mating preference for the recessive balances that for the dominant, giving the equilibrium frequency

$$q_* = \sqrt{[\theta/(\alpha + \theta)]}$$

This is identical to the equilibrium frequency of the previous section if $\theta = \gamma$.

In the general model combining preference for a dominant with separate preferences for each genotype, the difference equation can be written

$$\Delta q = -\tfrac{1}{2}(\alpha + \theta)(q - q_*)(q + q_* + a)/(1 + q)$$

where $a = \phi/(\alpha + \theta)$. As a differential equation, this has the solution

$$\left(\frac{q_T - q_*}{q_0 - q_*}\right)^{(q+1)} \cdot \left(\frac{q_T + q_* + a}{q_0 + q_* + a}\right)^{(q_* + a - 1)}$$
$$= \exp\left[-T(q_* + a)(\alpha + \theta)\right] \tag{4.2.6}$$

When preference for the dominant is balanced by preference for the recessive, $a = 0$, and the solution of the differential equation becomes identical to the solution obtained in section 4.1 putting $\theta = \gamma$ in equation 4.1.6.

4.3 Natural selection of dominant and recessive characters in males

In the most realistic model a character of sexual display is developed in the males improving their chances of mating, but at the same time lowering their fitness for survival. If such a character has become fixed in the population, its advantage in sexual selection must have outweighed its disadvantage in natural selection. In some cases, however, polymorphisms can exist: in certain populations of sticklebacks, some males do not develop the characteristic red-throat in the breeding season. Development of the red throat is inherited as a simple recessive. Although the red males are more successful in enticing the females to lay their eggs in the males' nests, in polymorphic populations they have been found to suffer very heavily from predation by trout. Like the females, the trout can presumably spot the red-throated males more easily. Elaborate sexual display must often entail such a disadvantage: sexual selection may thus balance natural selection at a polymorphic equilibrium. We may expect this to occur in the 'C' models in which the preferred males gain a frequency-dependent advantage in their mating success. As these males increase in frequency, their relative advantage declines. To increase in frequency at all, their initial mating advantage must exceed the disadvantage of any natural selection that may be acting against them. But if, as they increase in frequency, their mating advantage then declines, reaching a point at which it exactly balances natural selection, a stable polymorphism will be set up. The equilibrium must always be stable in these circumstances: if the preferred males were to increase still further in frequency their mating advantage would decline further and thus no longer balance the natural selection acting against the preferred males. Natural selection would therefore reduce their frequency, bringing the population back to the equilibrium point again.

If the preferred character is sex-limited to males, only they will be subject to the natural selection: the females will not be affected although they will carry the genotypes that determine the male characters. They mate with males of each genotype with the frequencies

$$P_T(AA) = \frac{u(1-\gamma)}{1-tw}$$

$$P_T(Aa) = \frac{v(1-\gamma)}{1-tw}$$

$$P_T(aa) = \gamma + \frac{w(1-\gamma)(1-t)}{1-tw}$$

$$(4.3.1)$$

when a recessive is the preferred character; and with the frequencies

$$P_T(AA) = \frac{\alpha u}{1-w} + \frac{u(1-\alpha)(1-s)}{1-s(1-w)}$$

$$P_T(Aa) = \frac{\alpha v}{1-w} + \frac{v(1-\alpha)(1-s)}{1-s(1-w)}$$

$$P_T(aa) = \frac{w(1-\alpha)}{1-s(1-w)}$$

$$(4.3.2)$$

when a dominant is preferred. Table 4.3 gives the frequencies of the matings in the two cases when either a recessive or a dominant character is the object of female preference. Thus for the recessive character we obtain

$$\Delta p = \tfrac{1}{2}p(-\gamma + tw)/(1 - tw)$$

$$(4.3.3)$$

and for the dominant

$$\Delta p = \tfrac{1}{2}pw[\alpha - s(1-w)]/[(1-w)(1-s+sw)]$$

$$(4.3.4)$$

These difference equations represent the change in gene frequency from the production of zygotes before natural selection had acted on them in one generation to the production of zygotes before natural selection in the next generation. At this point the preferred

Table 4.3 *Frequencies of matings when natural selection acts against a preferred character which is sex-limited to males*

Mating type	Frequency of mating type	
	Recessive character aa preferred	Dominant character AA or Aa preferred
$AA \times AA$	$\dfrac{u^2(1-\gamma)}{1-tw}$	$\dfrac{\alpha u^2}{1-w}+\dfrac{u^2(1-\alpha)(1-s)}{1-s(1-w)}$
$AA \times Aa$	$\dfrac{2uv(1-\gamma)}{1-tw}$	$\dfrac{2\alpha uv}{1-w}+\dfrac{2uv(1-\alpha)(1-s)}{1-s(1-w)}$
$AA \times aa$	$\gamma u+\dfrac{uw(1-\gamma)(2-t)}{1-tw}$	$\dfrac{\alpha uw}{1-w}+\dfrac{uw(1-\alpha)(2-s)}{1-s(1-w)}$
$Aa \times Aa$	$\dfrac{v^2(1-\gamma)}{1-tw}$	$\dfrac{\alpha v^2}{1-w}+\dfrac{v^2(1-\alpha)(1-s)}{1-s(1-w)}$
$Aa \times aa$	$\gamma v+\dfrac{vw(1-\gamma)(2-t)}{1-tw}$	$\dfrac{\alpha vw}{1-w}+\dfrac{vw(1-\alpha)(2-s)}{1-s(1-w)}$
$aa \times aa$	$\gamma w+\dfrac{w^2(1-\gamma)(1-t)}{1-tw}$	$\dfrac{w^2(1-\alpha)}{1-s(1-w)}$

t is the selective disadvantage of the recessive and s the selective disadvantage of the dominant. Thus, when the preferred character is recessive, males with genotypes AA, Aa and aa have frequencies $u/(1-tw)$, $v/(1-tw)$ and $w(1-t)/(1-tw)$ after natural selection. They mate with females with genotypic frequencies u, v and w of whom a proportion γ prefer aa males. When the preferred character is dominant, the males have frequencies $u(1-s)/[1-s(1-w)]$, $v(1-s)/[1-s(1-w)]$ and $w/[1-s(1-w)]$ after natural selection. A proportion α of the females prefer AA and Aa males indiscriminately.

phenotype has an equilibrium frequency of

$$w_* = \gamma/t \tag{4.3.5}$$

as a recessive and an equilibrium frequency of

$$1 - w_* = \alpha/s \tag{4.3.6}$$

as a dominant. If the mating preferences and natural selective coefficients are the same for both, the same phenotypic frequencies are therefore attained at equilibrium. At equilibrium the genotypes are maintained in the Hardy–Weinberg frequencies, $u_* = p_*^2$, $v_* = 2p_*q_*$ and $w_* = q_*^2$. Thus the equilibrium gene frequencies are given by

$$q_* = \sqrt{(\gamma/t)} \tag{4.3.7}$$

for the recessive allele and

$$p_* = 1 - \sqrt{(1 - \alpha/s)} \tag{4.3.8}$$

for the dominant. At the equilibrium of the recessive

$$d\Delta p/dp|_{p=p_*} = -p_* q_* t/(1 - t w_*)$$

showing that all equilibria are stable ($t > 0$). After natural selection of the males has taken place, they mate with an equilibrium frequency of $\gamma(1 - t)/t(1 - \gamma)$ for the recessive and $\alpha(1 - s)/s(1 - \alpha)$ for the dominant.

If the same intensities of selection act on dominant and recessive ($s = t$), we obtain the approximate difference equations

$$\Delta p = \tfrac{1}{2} sp(q^2 - q_*^2)$$

for the recessive and

$$\Delta p = \tfrac{1}{2} sq^2(q^2 - q_*^2)/(1 + q)$$

for the dominant.
These equations have the rather cumbrous solution for the recessive

$$\left(\frac{1 - q_0}{1 - q_T}\right)^{2q_*} \cdot \left(\frac{q_T - q_*}{q_0 - q_*}\right)^{(1 + q_*)} \cdot \left(\frac{q_0 + q_*}{q_T + q_*}\right)^{(1 - q_*)}$$
$$= \exp[-T s q_*(1 - q_*^2)] \tag{4.3.9}$$

and the even more cumbrous solution for the dominant

$$(1 + q_*)\log_e\left(\frac{q_T - q_*}{q_0 - q_*}\right) + (1 - q_*)\log_e\left(\frac{q_0 + q_*}{q_T + q_*}\right)$$
$$+ 2q_*\log_e\left(\frac{q_0}{q_T}\right) + \frac{2q_*}{q_T} - \frac{2q_*}{q_0} = -T s q_*^3 \tag{4.3.10}$$

These solutions, though accurate enough over a few generations, would be awkward to use for fitting data on gene frequencies and estimating parameters.

4.4   Natural selection of dominant and recessive characters in both sexes

If the preferred character is expressed in both males and females, natural selection will act on the genotypes in both sexes. The females as a whole still mate with frequencies given by equations 4.3.1 and

Table 4.4 *Frequencies of matings when natural selection acts against a preferred character expressed in both sexes*

Mating type	Recessive character *aa* preferred	Dominant character *AA* or *Aa* preferred
$AA \times AA$	$\dfrac{u^2(1-\gamma)}{(1-tw)^2}$	$\dfrac{\alpha u^2(1-s)}{(1-w)[1-s(1-w)]} + \dfrac{u^2(1-\alpha)(1-s)^2}{[1-s(1-w)]^2}$
$AA \times Aa$	$\dfrac{2uv(1-\gamma)}{(1-tw)^2}$	$\dfrac{2\alpha uv(1-s)}{(1-w[1-s(1-w)])} + \dfrac{2uv(1-\alpha)(1-s)^2}{[1-s(1-w)]^2}$
$AA \times aa$	$\dfrac{\gamma u}{1-tw} + \dfrac{2uw(1-\gamma)(1-t)}{(1-tw)^2}$	$\dfrac{\alpha uw}{(1-w)[1-s(1-w)]} + \dfrac{2uw(1-\alpha)(1-s)}{[1-s(1-w)]^2}$
$Aa \times Aa$	$\dfrac{v^2(1-\gamma)}{(1-tw)^2}$	$\dfrac{\alpha v^2(1-s)}{(1-w)[1-s(1-w)]} + \dfrac{v^2(1-\alpha)(1-s)^2}{[1-s(1-w)]^2}$
$Aa \times aa$	$\dfrac{\gamma v}{1-tw} + \dfrac{2uw(1-\gamma)(1-t)}{(1-tw)^2}$	$\dfrac{\alpha vw}{(1-w)[1-s(1-w)]} + \dfrac{2vw(1-\alpha)(1-s)}{[1-s(1-w)]^2}$
$aa \times aa$	$\dfrac{\gamma w}{1-tw} + \dfrac{w^2(1-\gamma)(1-t)^2}{(1-tw)^2}$	$\dfrac{w^2(1-\alpha)}{[1-s(1-w)]^2}$

The frequencies of the genotypes *AA*, *Aa* and *aa* for both sexes when they mate are now $u/(1-tw)$, $v/(1-tw)$ and $w(1-t)/(1-tw)$ for the recessive and $u(1-s)/[1-s(1-w)]$, $v(1-s)/[1-s(1-w)]$ and $w/[1-s(1-w)]$ for the dominant.

4.3.2 when the males are subject to natural selection; but the females now have the same genotypic frequencies after natural selection as the males. The frequencies of matings between the genotypes are then as shown in table 4.4.

If the recessive is the preferred character then it can easily be shown that

$$\Delta p = p(-\tfrac{1}{2}\gamma + tw)/(1-tw) \tag{4.4.1}$$

If the dominant is preferred, however, then after considerable algebraic simplification the expression for the change in gene frequency can be found as follows. The genotypes *AA* and *Aa* have zygotic frequencies in the next generation of

$$u' = \frac{\alpha p^2(1-s)}{(1-w)[1-s(1-w)]} + \frac{p^2(1-\alpha)(1-s)^2}{[1-s(1-w)]^2}$$

$$v' = \frac{\alpha pv(1-s) + \alpha pw}{(1-w)[1-s(1-w)]}$$
$$+ \frac{2pq(1-\alpha)(1-s)^2 + 2spw(1-\alpha)(1-s)}{[1-s(1-w)]^2}$$

Therefore

$$p' = u' + \tfrac{1}{2}v'$$

$$= \frac{\alpha p(1-s)(1-\tfrac{1}{2}w) + \tfrac{1}{2}\alpha spw}{(1-w)[1-s(1-w)]} + \frac{p(1-\alpha)(1-s)}{[1-s(1-w)]}$$

and

$$\Delta p = \frac{pw[\tfrac{1}{2}\alpha - s(1-w)]}{(1-w)[1-s(1-w)]} \qquad (4.4.2)$$

So we have the equilibrium frequencies of the preferred phenotypes

$$w_* = \tfrac{1}{2}\gamma/t \quad \text{(recessive)} \qquad (4.4.3)$$

$$1 - w_* = \tfrac{1}{2}\alpha/s \quad \text{(dominant)} \qquad (4.4.4)$$

As we should expect when natural selection acts in both sexes, the mating preference must be doubled to produce the same equilibrium frequency as when natural selection acts only in males (equations 4.3.5 and 4.3.6). The rate of approach to a given equilibrium frequency is also doubled: for the recessive, the rate depends on the expression $\exp[-2Tsq_*(1-q_*^2)]$ compared to $\exp[-Tsq_*(1-q_*^2)]$ when natural selection acts only in males (equation 4.3.9). The polymorphic equilibria are easily shown to be stable.

4.5   Fitting the models to data

In many experiments on mating success in *Drosophila*, males with rare genotypes have had the sexual advantage (Ehrman, 1967, 1968; Spiess, 1968; Ehrman and Spiess, 1969; Spiess and Spiess, 1969). Ehrman and Spiess assumed that female preferences or differences in male courtship must therefore depend on the relative frequency of the male genotypes. However, their observations fit the model of constant preferences for different phenotypes: although the preferences are constant, the mating advantage is frequency-dependent. Thus if α of the females prefer males A and γ prefer males B, then males A and B should mate in proportions $\alpha + x(1-\alpha-\gamma)$ and $\gamma + y(1-\alpha-\gamma)$ where x and y are the frequencies of A and B. Spiess (1968) and Spiess and Spiess (1969) carried out some carefully controlled experiments in which males with varying frequencies of two genotypes were allowed to mate with females with equal frequencies of the two genotypes. The genotypic frequencies of the males were varied between experiments, taking the nine values 0.1, 0.2, ..., 0.9. In one experiment (Spiess, 1968), the genotypes were karyotypes for two chromosomal inversions AR and PP in *Drosophila pseudoobscura* sampled from the same locality. In another experiment (Spiess and Spiess, 1969) the genotypes were homokaryotypic strains of *Drosophila persimilis* taken from two different localities Humboldt (Hu) and White Wolf (WW). Maximum likelihood estimates of female mating preferences, α and γ, have been obtained as follows (O'Donald, 1977a). They were found by trial and error, starting with arbitrarily chosen values of α and γ and introducing random perturbations into these values until higher values of the log likelihood were obtained. By making the random perturbations smaller and smaller, the maximum likelihood is thus reached by a random walk of successively smaller steps.

(i) In the experiment with *Drosophila pseudoobscura*, according to the model,
$\hat{\alpha} = 0.138$ for females preferring AR males
$\hat{\gamma} = 0.021$ for females preferring PP males

Before fitting the parameters α and γ, $\chi_9^2 = 67.105$ for the nine degrees of freedom in deviations of numbers of matings from expectations calculated according to male frequency. After fitting the parameters $\chi_7^2 = 8.672$ $(0.5 > P > 0.3)$ for the residual heterogeneity.

(ii) In the experiment with *Drosophila persimilis*,
$\hat{\alpha} = 0.166$ for females preferring Hu males
$\hat{\gamma} = 0.156$ for females preferring WW males

Before fitting parameters, $\chi_9^2 = 101.707$. After fitting $\chi_7^2 = 13.692$ $(0.1 > P > 0.05)$.

This result agrees with the general hypothesis that mating preferences evolve by the selection of those females that choose selectively advantageous males who are both fitter individuals and who will produce fitter offspring. Disruptive selection will produce preferences for males of different strains: in Thoday's experiments on disruptive selection in *Drosophila* (Thoday and Gibson, 1962), mating preferences evolved rapidly for phenotypes selected with high or low bristle numbers; assortment also evolved giving rise to sexual isolation.

As well as fitting data of numbers of matings to the model, data of the gene frequencies in successive generations could also be fitted, thus giving independent estimates of the parameters. The final equilibrium frequency can also be compared with the frequency predicted from the estimates of the preferences. Thus if the preferences we have estimated for *AR* and *PP* males in Spiess' experiments are preferences just for the separate homozygous genotypes, we should expect that the equilibrium frequency of the inversion *AR* will be given by equation 4.2.2

$$p_* = 0.138/0.159$$

$$= 0.868$$

On the other hand, if these were preferences for *AR* homozygotes or heterozygotes and *PP* homozygotes, then from equation 4.1.5

$$p_* = 1 - \sqrt{(0.021/0.159)}$$

$$= 0.637$$

Equilibrium frequencies of these karyotypes in population cage experiments can therefore be used for an independent test of the model. Since heterozygotes were not tested for mating preferences against the homozygotes, it would be arbitrary to assume that mating preferences may operate in their favour. Computer simulation (see Chapter 8, section 8.3) has shown that separate preferences for heterozygotes are usually eliminated while preferences for homozygotes increase in frequency. If so, the preferences we have estimated would only operate in favour of the homozygotes tested in the experiments. Then we should expect the frequency of the inversion *AR* to be given by equation 4.2.2 so that we may predict

$$p_* = 0.868$$

Predictions based on the 'C', 'P' and 'PF −' models of the expression of mating preferences are compared in Chapter 6, section 6.9. All the models make very similar predictions of the equilibrium gene frequency p_*.

Ehrman and Spiess' experiments do not prove that females differ in their mating preferences. Their results strongly support the theory of sexual selection by female preference, however, since the males mate with a negative frequency-dependence that is to be expected if some females prefer one phenotype of male and other females prefer another phenotype; differences in the males' competitive ability, on the contrary, would give rise to slight positive frequency-dependence. Semler (1971) gave a direct experimental proof that females exercise preferences in his experiment on stickle-backs (see section 1.1) by showing that the females always preferred to lay their eggs in the nests of red-throated males regardless of whether the male was either genetically red-throated or non-red with an artificial red throat painted on with lipstick. Tebb and Thoday (1956) demonstrated that female genotype can determine preferential mating for two different stocks of *Drosophila melanogaster*: homozygous females preferred males of one stock; heterozygous females preferred the others. The homozygous and heterozygous females thus differed in their mating responses to the males. Tebb and Thoday used stocks marked with the sex-linked mutants *white eye* (w) or *apricot eye* (w^a) in their experiment. Females, genotypically either w/w, w/w^a or w^a/w^a, were given a choice of mating with either w or w^a males. Table 4.5 shows the data and my analysis of the components of the data that contribute to the total value of χ^2. The preferences of homozygous females for w^a males and heterozygous females for w males are highly significant, the values of χ^2 corresponding to probabilities $P < 0.01$. No other effect is statistically significant.

Suppose α is the preference of the homozygotes and β the preference of the heterozygotes. Then, after combining the results of the two experiments at different temperatures, the matings should take place with the following probabilities:

w/w		w^a/w^a		w/w^a	
w	w^a	w	w^a	w	w^a
$\frac{1}{6}(1-\alpha)$	$\frac{1}{6}(1+\alpha)$	$\frac{1}{6}(1-\alpha)$	$\frac{1}{6}(1+\alpha)$	$\frac{1}{6}(1+\beta)$	$\frac{1}{6}(1-\beta)$

The independent linear component of the observations corresponding to the preference α has a theoretical value

$$\frac{n}{6}[(1-\alpha)-(1+\alpha)+(1-\alpha)-(1+\alpha)] = -2n\alpha/3$$

and an observed value

$$7-19+14.5-14.5+10-17+9-19 = -29$$

Hence, on equating the theoretical and observed values, we get the

Table 4.5 Analysis of χ^2 of Tebb and Thoday's data on preferential matings of white eye and apricot eye in Drosophila melanogaster

Factor	25°C						15–25°C						Value of χ^2
	w/w		w^a/w^a		w/w^a		w/w		w^a/w^a		w/w^a		
	w	w^a	w	w^a	w	w^a	w	w^a	w	w^a	w	w^a	
—	7	19	14.5	14.5	18	10	10	17	9	19	21.5	8.5	19.1428
T	1	1	1	1	1	1	−1	−1	−1	−1	−1	−1	0.0238
f_A	1	1	−1	−1	0	0	1	1	−1	−1	0	0	0.1429
f_D	1	1	1	1	−2	−2	1	1	1	1	−2	−2	0.1072
α	1	−1	1	−1	0	0	1	−1	1	−1	0	0	7.5089
β	0	0	0	0	1	−1	0	0	0	0	1	−1	7.8750
$f_A \times \alpha$	1	−1	−1	1	0	0	1	−1	−1	1	0	0	0.7232
$T \times f_A$	1	1	−1	−1	0	0	−1	−1	1	1	0	0	0.0357
$T \times f_D$	1	1	1	1	−2	−2	−1	−1	−1	−1	2	2	0.0476
$T \times \alpha$	1	−1	1	−1	0	0	−1	1	−1	1	0	0	0.2232
$T \times \beta$	0	0	0	0	1	−1	0	0	0	0	−1	1	0.4464
$T \times f \times \alpha$	1	−1	−1	1	0	0	−1	1	1	−1	0	0	2.0089

The values shown in the first row of the table are the numbers of matings of white eyed (w) and apricot eyed (w^a) males observed in Tebb and Thoday's experiment; the value of χ^2 has 11 degrees of freedom for deviations from the expectation of equal numbers. The remaining rows give the coefficients of linear functions of the observations for eleven independent factors and the values of χ^2 for 1 degree of freedom corresponding to each factor as follows:

T is the effect of difference in temperature;

f_A is the additive effect of female genotypes measured by the difference in the matings of females who are genotypically w/w or w^a/w^a; f_D is the dominance effect of female genotypes measured by the difference between females who are homozygous or heterozygous; α is the preference of homozygous females for w^a males; β is the preference of heterozygous females for w males; and the remaining factors are the interactions between these effects.

estimate

$$\hat{\alpha} = 0.2589$$

According to the null hypothesis $\alpha = 0$, the linear component has a variance $2n/3 = 112$ giving

$$\chi^2 = (-29)^2/112$$

$$= 7.5089$$

as shown in the table. This corresponds to a level of significance given by the probability

$$P = 0.00614$$

Similarly, for the independent linear component corresponding to the preference β, we equate the theoretical value $n\beta/3$ to the observed value 21 to obtain the estimate

$$\hat{\beta} = 0.375$$

with

$$\chi^2 = 7.875$$

and

$$P = 0.00501$$

These estimates of α and β are nearly but not quite the same as the following estimates obtained by maximum likelihood:

$$\hat{\alpha} = 0.2636$$

$$\hat{\beta} = 0.3621$$

Since the males did not differ in their overall chances of mating, the preferential matings were solely determined by the genotypes of the females who exercised the choice. Sexual selection took place because females differed in their preferences.

5

Partial expression of mating preference

5.1 Preferences for dominant and recessive characters

In the 'P' models with partial expression of the mating preferences, I showed in section 3.2 that the frequencies of matings with males who are phenotypically A with frequency x or B with frequency y are given by equations 3.2.1

$$P_T(A) = \alpha(1 - y^{n+1}) + \gamma x^{n+1} + x(1 - \alpha - \gamma)$$

$$P_T(B) = \alpha' y^{n+1} + \gamma(1 - x^{n+1}) + y(1 - \alpha - \gamma)$$

As in the subsequent analysis of the corresponding models of complete mating preference, we then find the frequencies of matings with each of the genotypes AA, Aa and aa on the assumption that A is dominant (genotypically AA or Aa) and B is recessive (aa). This gives the equations

$$
\left.
\begin{aligned}
P_T(AA) &= \alpha\left(\frac{u}{1-w}\right)(1 - w^{n+1}) + \gamma\left(\frac{u}{1-w}\right)(1 - w)^{n+1} \\
&\quad + u(1 - \alpha - \gamma) \\
P_T(Aa) &= \alpha\left(\frac{v}{1-w}\right)(1 - w^{n+1}) + \gamma\left(\frac{v}{1-w}\right)(1 - w)^{n+1} \\
&\quad + v(1 - \alpha - \gamma) \\
P_T(aa) &= \alpha w^{n+1} + \gamma[1 - (1 - w)^{n+1}] + w(1 - \alpha - \gamma)
\end{aligned}
\right\} \quad (5.1.1)
$$

Since natural selection is not assumed to be taking place in the model analysed in this section, these are the frequencies at which females mate with each of the male genotypes. The females are also genotypically AA, Aa or aa with frequencies u, v and w.

According to the general derivation of the 'P' models, the parameter n may be regarded either as the number of extra encounters a female must make with those males who are not the object of her preference before she is willing to mate at random with any male, or $n+1$ is the number of males in a group from which a female can make her choice of male. Mathematical solutions giving equilibrium frequencies in these 'P' models seem to be obtainable only in the cases when $n=1$ or $n=2$. Thus, if $n=1$, one extra encounter with a courting male will be sufficient to raise a female's level of sexual stimulation from the threshold at which she will respond only to the males she prefers up to the threshold at which she will respond to any male: thus the difference in the two thresholds cannot be greater than the stimulus provided by a single encounter with a courting male. If $n=2$, the difference in the thresholds must be greater than the stimulus provided by one encounter but less than the stimulus provided by two encounters. When $n>2$, numerical computation must be used to investigate the equilibria of the model. As n increases, the equilibrium frequencies should converge on those of the corresponding 'C' model which represents the limiting case of the 'P' model as $n \to \infty$.

When females need only one or two extra encounters before mating at random, the models have simple mathematical solutions. The frequencies of matings according to equations 5.1.1 with which females mate with males of each genotype, then become the following:

$$\left.\begin{aligned}
P_T(AA) &= u[1+w(\alpha-\gamma)] \\
P_T(Aa) &= v[1+w(\alpha-\gamma)] \\
P_T(aa) &= w[1+w(\alpha-\gamma)] - w(\alpha-\gamma)
\end{aligned}\right\} \qquad (5.1.2)$$

for the case when $n=1$; and

$$\left.\begin{aligned}
P_T(AA) &= u\{1+w[\alpha-2\gamma+w(\alpha+\gamma)]\} \\
P_T(Aa) &= v\{1+w[\alpha-2\gamma+w(\alpha+\gamma)]\} \\
P_T(aa) &= w\{1-(1-w)[\alpha-2\gamma+w(\alpha+\gamma)]\}
\end{aligned}\right\} \qquad (5.1.3)$$

for the case when $n=2$.

Table 5.1 then gives the frequencies of the different mating types in these cases. When $n=1$, the recurrence equations for the genotypic frequencies in the next generation are as follows:

$$u' = p^2[1+w(\alpha-\gamma)]$$
$$v' = 2pq[1+w(\alpha-\gamma)] - pw(\alpha-\gamma)$$
$$w' = q^2[1+w(\alpha-\gamma)] - qw(\alpha-\gamma)$$

Table 5.1   *Frequencies of mating types with a dominant gene when one or two extra encounters with males are necessary before females with preferences mate at random*

Mating type	Frequency of mating type	
	With one encounter before random mating $n=1$	With two encounters before random mating $n=2$
$AA \times AA$	$u^2[1+w(\alpha-\gamma)]$	$u^2\{1+w[\alpha-2\gamma+w(\alpha+\gamma)]\}$
$AA \times Aa$	$2uv[1+w(\alpha-\gamma)]$	$2uv\{1+w[\alpha-2\gamma+w(\alpha+\gamma)]\}$
$AA \times aa$	$uw[2+(2w-1)(\alpha-\gamma)]$	$uw\{2+(2w-1)[\alpha-2\gamma+w(\alpha+\gamma)]\}$
$Aa \times Aa$	$v^2[1+w(\alpha-\gamma)]$	$v^2\{1+w[\alpha-2\gamma+w(\alpha+\gamma)]\}$
$Aa \times aa$	$vw[2+(2w-1)(\alpha-\gamma)]$	$vw\{2+(2w-1)[\alpha-2\gamma+w(\alpha+\gamma)]\}$
$aa \times aa$	$w^2[1+(w-1)(\alpha-\gamma)]$	$w^2\{1+(w-1)[\alpha-2\gamma+w(\alpha+\gamma)]\}$

so that

$$\Delta p = \tfrac{1}{2}pw(\alpha-\gamma) \tag{5.1.4}$$

If $\alpha>\gamma$, $p_T\to1$: if $\gamma>\alpha$, $p_T\to0$. No polymorphism can exist if $n=1$.

Approximately, putting $w=q^2$ we have the difference equation $\Delta p=\tfrac{1}{2}pq^2(\alpha-\gamma)$ and hence the change in gene frequency from p_0 to p_T is given by

$$\tfrac{1}{2}T(\alpha-\gamma)=\log_e\left(\frac{p_Tq_0}{p_0q_T}\right)+\frac{1}{q_T}-\frac{1}{q_0}$$

When $n=2$, the corresponding recurrence equations are then as follows:

$$u' = p^2\{1+w[\alpha-2\gamma+w(\alpha+\gamma)]\}$$
$$v' = 2pq\{1+w[\alpha-2\gamma+w(\alpha+\gamma)]\}-pw[\alpha-2\gamma+w(\alpha+\gamma)]$$
$$w' = q^2\{1+w[\alpha-2\gamma+w(\alpha+\gamma)]\}-qw[\alpha-2\gamma+w(\alpha+\gamma)]$$

so that

$$\Delta p = \tfrac{1}{2}pw[\alpha-2\gamma+w(\alpha+\gamma)] \tag{5.1.5}$$

At equilibrium $u_*=p_*^2$, $v_*=2p_*q_*$ and $w_*=(2\gamma-\alpha)/(\alpha+\gamma)=q_*^2$ so that

$$q_*=\sqrt{[(2\gamma-\alpha)/(\alpha+\gamma)]}$$

Approximately therefore

$$\Delta p = -\tfrac{1}{2}(\alpha+\gamma)q^2(1-q)(q^2-q_*^2)$$

Differentiating, we find there exists a stable polymorphic equilibrium if $2\gamma>\alpha$.

Table 5.2   *Genetic equilibria with mating preferences for dominant and recessive genes and different numbers of female encounters before mating*

Preference for dominant phenotype determined by AA and Aa genotypes: $\alpha = 0.1$

No. of extra encounters n	Values of preference, γ, for aa males			
	0.05	0.1	0.15	0.2
1	1.0	*	0	0
2	1.0	0.293	0.106	0
3	0.561	0.293	0.176	0.101
4	0.490	0.293	0.201	0.140
5	0.460	0.293	0.212	0.158
10	0.426	0.293	0.224	0.181
20	0.423	0.293	0.225	0.183
30	0.423	0.293	0.225	0.183
∞	0.423	0.293	0.225	0.183

The symbol * indicates a neutrally stable point. All other equilibria are globally stable (O'Donald, 1978a).

Generally for values of $n > 5$, the equilibria converge closely on those of the model with complete preferences. This is shown in Table 5.2 for particular values of α and γ. If $\alpha = \gamma$ and $n = 1$ the equilibrium is neutrally stable since $\Delta p = 0$ for all values of p. For the model with complete preferences ($n \to \infty$) the equilibrium frequency is $p_* = 1 - \sqrt{[\gamma/(\alpha + \gamma)]}$ and always stable. For any value of n, the value of w_* may be found as one of the roots of the polynomial

$$[\alpha + (-1)^n \gamma]w^n + (-1)^{n-1}\gamma(n+1)w^{n-1}$$
$$+ (-1)^{n-2}[\gamma n(n+1)/2]w^{n-2} + \ldots$$
$$+ [\gamma n(n+1)(n-1)/3!]w^2 - [\gamma n(n+1)/2]w$$
$$+ n\gamma - \alpha = 0$$

The existence of stable polymorphic equilibria for the case when $n > 1$ is a consequence of the negative frequency-dependence of the mating advantage which the males gain as a result of female preference. But for the case when $n = 1$, the frequency-dependence is slightly positive. For example, consider the relative mating success of the two phenotypes A and B with preferences α and γ in their favour in the case when $n = 1$.

Then we have

$$P_T(A) = x[1 + y(\alpha - \gamma)]$$
$$P_T(B) = y[1 - x(\alpha - \gamma)]$$

Since A and B occur at frequencies x and y, the relative mating advantage of B is

$$[1 - x(\alpha - \gamma)]/(1 + y(\alpha - \gamma))]$$

Hence the selective coefficient of B relative to A is

$$1 - [1 - x(\alpha - \gamma)]/[1 + y(\alpha - \gamma)] = (\alpha - \gamma)/[1 + y(\alpha - \gamma)]$$

If $\alpha > \gamma$ so that A has the overall advantage, then the relative disadvantage of B is $(\alpha - \gamma)$ when rare and $(\alpha - \gamma)/(1 + \alpha - \gamma)$ when on the point of fixation. Its disadvantage is thus slightly greater when it is rare; correspondingly the advantage of A increases slightly as A becomes common. The phenotype that has the initial advantage slightly increases its advantage as it spreads through the population to ultimate fixation. If $n = 2$, however, A would have an initial selective advantage of approximately $(2\alpha - \gamma)/(1 + 2\alpha - \gamma)$ which would decline to $\alpha - 2\gamma$ when A was on the point of fixation. Thus provided $2\gamma > \alpha > \frac{1}{2}\gamma$, which is the condition for the existence and stability of the polymorphic equilibrium, A will start at an advantage, but this will decline to a point at which it will be balanced by the increasing advantage of the B phenotype: in general the selective coefficient of B is equal to $[\alpha - 2\gamma + y(\alpha + \gamma)]/\{1 + y[\alpha - 2\gamma + y(\alpha + \gamma)]\}$ which equals zero when $y = (2\gamma - \alpha)/(\alpha + \gamma)$. This is the stable equilibrium frequency of the B phenotype, giving $q_* = \sqrt{[(2\gamma - \alpha)/(\alpha + \gamma)]}$ as we have shown when B is the recessive.

5.1.1   *Mating preferences with different thresholds in female response*

Females may vary in the thresholds at which they express their different preferences: a preference for one phenotype may be expressed at a lower threshold than a preference for the other. If so, females who prefer one or other of the phenotypes would therefore differ in the number of extra encounters they make before mating at random among the males. Suppose, as suggested in section 3.2, that the males mate with the following frequencies:

$$P_T(AA) = \alpha\left(\frac{u}{1-w}\right)(1 - w^{n+1}) + \gamma\left(\frac{u}{1-w}\right)(1 - w)^{m+1}$$
$$+ u(1 - \alpha - \gamma)$$

$$P_T(Aa) = \alpha\left(\frac{v}{1-w}\right)(1-w^{n+1}) + \gamma\left(\frac{v}{1-w}\right)(1-w)^{m+1}$$
$$+ v(1-\alpha-\gamma)$$

$$P_T(aa) = \alpha w^{n+1} + \gamma[1-(1-w)^{m+1}] + w(1-\alpha-\gamma)$$

Such a model implies that females with a preference for the dominant A phenotype have a different lower threshold from females with a preference for the recessive a phenotype. Different numbers of encounters, n and m, will therefore be required before females with preferences for each of these males mate at random.

The following cases of this model lead to explicit mathematical results: (i) $n=1$ and $m=2$; (ii) $n=2$ and $m=1$; (iii) $n=1$ and $m \to \infty$; (iv) $n \to \infty$ and $m=1$.

(i) *Number of encounters $n=1$ and $m=2$*

The frequencies of matings are given by

$$P_T(AA) = u(1+\alpha w - 2\gamma w + \gamma w^2)$$
$$P_T(Aa) = v(1+\alpha w - 2\gamma w + \gamma w^2)$$
$$P_T(aa) = w(1+\alpha w - 2\gamma w + \gamma w^2) - w(\alpha - 2\gamma + \gamma w)$$

giving recurrence equations for the genotypic frequencies

$$u' = p^2(1+\alpha w - 2\gamma w + \gamma w^2)$$
$$v' = 2pq(1+\alpha w - 2\gamma w + \gamma w^2) - pw(\alpha - 2\gamma + \gamma w)$$
$$w' = q^2(1+\alpha w - 2\gamma w + \gamma w^2) - qw(\alpha - 2\gamma + \gamma w)$$

and hence

$$p' = p(1+\tfrac{1}{2}\alpha w - \gamma w + \tfrac{1}{2}\gamma w^2)$$
$$\Delta p = \tfrac{1}{2}pw(\alpha - 2\gamma + \gamma w)$$

Therefore an equilibrium occurs at the frequencies

$$w_* = (2\gamma - \alpha)/\gamma \qquad 1 - w_* = (\alpha - \gamma)/\gamma$$

At this equilibrium we find

$$u_* = p_*^2$$
$$v_* = 2p_*q_*$$
$$w_* = q_*^2$$

showing that the genotypes are maintained at Hardy–Weinberg frequencies.

Since we may write

$$\Delta p = \tfrac{1}{2}\gamma p w (w - w_*)$$

we can then show that at the equilibrium frequency

$$d\Delta p/dp|_{p=p_*} = -\gamma p_* q_* w_*$$

Therefore a polymorphism exists if $\alpha > \gamma$. It is stable if $\gamma > 0$.

It is interesting that although no stable polymorphism can exist if $n = m = 1$, we now find one if $n = 1$ and $m = 2$. It exists because the a phenotype has a frequency-dependent advantage, whereas A does not. Provided $\alpha > \gamma$, the allele A can increase in frequency. But as it does so, the advantage of a increases and a balance can eventually be reached between the roughly constant advantage of A and the frequency-dependent advantage of a.

(ii) *Number of encounters $n = 2$ and $m = 1$*

The frequencies of matings are now given by

$$P_T(AA) = u(1 + \alpha w + \alpha w^2 - \gamma w)$$

$$P_T(Aa) = v(1 + \alpha w + \alpha w^2 - \gamma w)$$

$$P_T(aa) = w(1 + \alpha w + \alpha w^2 - \gamma w) + w(\alpha + \alpha w - \gamma)$$

giving the recurrence equation

$$p' = p(1 + \tfrac{1}{2}\alpha w + \tfrac{1}{2}\alpha w^2 - \tfrac{1}{2}\gamma w)$$

At equilibrium we have phenotypic frequencies

$$w_* = (\gamma - \alpha)/\alpha \qquad 1 - w_* = (2\alpha - \gamma)/\alpha$$

The genotypes are again maintained at the Hardy–Weinberg frequencies. The equilibrium exists if $\gamma > \alpha$ and is stable if $\alpha > 0$.

(iii) *Number of encounters $n = 1$ and $m \to \infty$*

In this case the preference for the a phenotype is complete and does not depend on frequency. Thus we have the following frequencies of mating:

$$P_T(AA) = u(1 + \alpha w - \gamma)$$

$$P_T(Aa) = v(1 + \alpha w - \gamma)$$

$$P_T(aa) = w(1 + \alpha w - \gamma) - \alpha w + \gamma$$

and therefore

$$p' = p(1 + \tfrac{1}{2}\alpha w - \tfrac{1}{2}\gamma)$$

giving the equilibrium

$$w_* = \gamma/\alpha$$

which exists and is stable if $\alpha > \gamma > 0$.

(iv) *Number of encounters $n \to \infty$ and $m = 1$*

Now the preference for the A phenotype is complete. We have the frequencies of matings

$$P_T(AA) = \alpha u/(1-w) + u(1-\alpha-\gamma w)$$

$$P_T(Aa) = \alpha v/(1-w) + v(1-\alpha-\gamma w)$$

$$P_T(aa) = \gamma w + w(1-\alpha-\gamma w)$$

giving

$$\Delta p = \tfrac{1}{2} p w [\alpha - \gamma(1-w)]/(1-w)$$

so that at equilibrium

$$1 - w_* = \alpha/\gamma$$

which exists and is stable if

$$\gamma > \alpha > 0$$

5.2. Separate preferences for each genotype

When females prefer particular genotypes, the parameters α, β and γ represent proportions of females with preferences for genotypes AA, Aa and aa; this is therefore a model with no dominance of the preferred genotypes. The females then choose to mate with males who are genotypically AA, Aa or aa with frequencies

$$P_T(AA) = \alpha[1 - (1-u)^{n+1}] + \beta\left(\frac{u}{u+w}\right)(1-v)^{n+1}$$

$$+ \gamma\left(\frac{u}{u+v}\right)(1-w)^{n+1} + \delta u$$

$$P_T(Aa) = \alpha\left(\frac{v}{v+w}\right)(1-u)^{n+1} + \beta[1 - (1-v)^{n+1}]$$

$$+ \gamma\left(\frac{v}{u+v}\right)(1-w)^{n+1} + \delta v$$

$$(5.2.1)$$

$$P_T(aa) = \alpha\left(\frac{w}{v+w}\right)(1-u)^{n+1} + \beta\left(\frac{w}{u+w}\right)(1-v)^{n+1}$$

$$+ \gamma[1 - (1-w)^{n+1}] + \delta w \Bigg\}$$

where

$$\delta = 1 - \alpha - \beta - \gamma$$

Three special cases of this model have analytical solutions: $n = 1$, $n = 2$ and $n \to \infty$. As $n \to \infty$, the model simply reduces to the 'C' models of sexual selection in which females' preferences are always expressed.

Table 5.3 gives the frequencies of the different mating types when a female must encounter either one or two extra males before mating at random among all the males and not just the males she prefers. Then, when $n = 1$, we have the following recurrence equations giving the frequencies, u', v', and w', of the genotypes AA, Aa and aa in the next generation.

$$u' = p^2(1 - \alpha u - \beta v - \gamma w) + p(\alpha u + \tfrac{1}{2}\beta v)$$

$$v' = 2pq(1 - \alpha u - \beta v - \gamma w) + q(\alpha u + \tfrac{1}{2}\beta v) + p(\gamma w + \tfrac{1}{2}\beta v)$$

$$w' = q^2(1 - \alpha u - \beta v - \gamma w) + q(\gamma w + \tfrac{1}{2}\beta v)$$

Table 5.3 *Frequencies of mating types with separate preferences for the genotypes when one or two extra encounters with males are necessary before females with preferences mate at random*

	Frequency of mating type	
Mating type	With one extra encounter before random mating $n = 1$	With two extra encounters before random mating $n = 2$
$AA \times AA$	$u^2(1 + \alpha - \eta)$	$u^2(1 + 2\alpha - \alpha u - 2\eta + \epsilon)$
$AA \times Aa$	$uv(2 + \alpha + \beta - 2\eta)$	$uv(2 + 2\alpha + 2\beta - \alpha u - \beta u - 4\eta + 2\epsilon)$
$AA \times aa$	$uw(2 + \alpha + \gamma - 2\eta)$	$uw(2 + 2\alpha + 2\gamma - \alpha u - \gamma w - 4\eta + 2\epsilon)$
$Aa \times Aa$	$v^2(1 + \beta - \eta)$	$v^2(1 + 2\beta - \beta v - 2\eta + \epsilon)$
$Aa \times aa$	$vw(2 + \beta + \gamma - 2\eta)$	$vw(2 + 2\beta + 2\gamma - \beta v - \gamma w - 4\eta + 2\epsilon)$
$aa \times aa$	$w^2(1 + \gamma - \eta)$	$w^2(1 + 2\gamma - \gamma w - 2\eta + \epsilon)$
	$\eta = \alpha u + \beta v + \gamma w$	$\epsilon = \alpha u^2 + \beta v^2 + \gamma w^2$

where

$$p = u + \tfrac{1}{2}v \quad \text{and} \quad q = w + \tfrac{1}{2}v.$$

At equilibrium we have the equation $p = (\alpha u + \tfrac{1}{2}\beta v)/(\alpha u + \beta v + \gamma w)$ and therefore $u_* = p_*^2$, $v_* = 2p_* q_*$ and $w_* = q_*^2$ giving $p_* = (\gamma - \beta)/(\alpha - 2\beta + \gamma)$ which is the equilibrium gene frequency at which a polymorphism may be established. If $\alpha = \beta = \gamma$, p_* is not defined, and the recurrence equations reduce simply to those of the Hardy–Weinberg Law.

It can be shown that the equations $u = p^2$, $v = 2pq$ and $w = q^2$ are a good approximation during the approach to equilibrium. Substitution of these values in the general recurrence equations gives the difference equation

$$\Delta p = \tfrac{1}{2}pq[p(\alpha - 2\beta + \gamma) - (\gamma - \beta)]$$

$$= \tfrac{1}{2}(\alpha - 2\beta + \gamma)pq(p - p_*) \tag{5.2.2}$$

Differentiating this equation then gives the condition for the stability of the equilibrium at p_*; for we find

$$d\Delta p/dp \big|_{p = p_*} = \tfrac{1}{2}(\alpha - 2\beta + \gamma)p_* q_*$$

and for the local stability of p_*

$$d\Delta p/dp \big|_{p = p_*} < 0$$

so that the equilibrium is always stable provided $\beta > \max(\alpha, \gamma)$. Stability thus requires heterosis in the proportion of females expressing preferences: the equilibrium is stable provided the proportion of females with a preference for the heterozygotes is greater than the average of those that prefer homozygotes.

The equation giving the polymorphic equilibrium may be compared with that in the model for complete preferences. Then we have the equilibrium gene frequency

$$p_* = (\alpha + \tfrac{1}{2}\beta)/(\alpha + \beta + \gamma)$$

which is always stable. Table 5.4 shows some special cases.

The general recurrence equation 5.2.2. can be integrated over a period of T generations during which the gene frequency changes from p_0 to p_T. The solution is found to be

$$\left(\frac{p_T - p_*}{p_0 - p_*}\right)\left(\frac{p_0}{p_T}\right)^{q_*}\left(\frac{q_0}{q_T}\right)^{p_*} = \exp[\tfrac{1}{2}Tp_* q_*(\alpha - 2\beta + \gamma)] \tag{5.2.3}$$

Table 5.4   *Equilibria and their stability for some special cases*

Special case	Model when $n=1$		Model when $n \to \infty$	
	Equilibrium frequency	Stability	Equilibrium frequency	Stability
$\alpha = 0$	$(\beta - \gamma)/(2\beta - \gamma)$	Stable if $\beta > \gamma$	$\beta/(2\beta + 2\gamma)$	Stable
$\beta = 0$	$\gamma/(\alpha + \gamma)$	Unstable	$\alpha/(\alpha + \gamma)$	Stable
$\gamma = 0$	$\beta/(2\beta - \alpha)$	Stable if $\beta > \alpha$	$(\alpha + \frac{1}{2}\beta)/(\alpha + \beta)$	Stable
$\alpha = \beta = 0$	1	Unstable	0	Stable
$\alpha = \gamma = 0$	$\frac{1}{2}$	Stable	$\frac{1}{2}$	Stable
$\beta = \gamma = 0$	0	Unstable	1	Stable

The general absence of stability in the model when $n=1$, except in cases of heterotic mating preference, is a consequence of positive frequency-dependence in the mating advantage gained by the males (O'Donald, 1978a).

where $q_* = 1 - p_*$. Obviously this solution is valid only when $\beta > \frac{1}{2}(\alpha + \gamma)$. When $\alpha = \beta = 0$, $p_* = 1$; but since

$$\Delta p = -\frac{1}{2} p \gamma w$$

$$\cong -\frac{1}{2}\gamma p q^2$$

therefore $p_T \to 0$ as we should expect, and to a close approximation

$$\log_e \left(\frac{p_0 \, q_T}{p_T \, q_0} \right) - \frac{1}{q_T} + \frac{1}{p_0} = \frac{1}{2}\gamma T$$

When $\gamma = \beta = 0$, $p_* = 0$; but $\Delta p = \frac{1}{2}\alpha p^2 q$, $p_T \to 1$, and

$$\log_e \left(\frac{p_T \, q_0}{p_0 \, q_T} \right) - \frac{1}{p_T} + \frac{1}{p_0} = \frac{1}{2}\alpha T$$

If females with preferences need two extra encounters before mating with males who are not of their choice ($n=2$), then the recurrence equations are

$$u' = p^2[1 - 2(\alpha u + \beta v + \gamma w) + (\alpha u^2 + \beta v^2 + \gamma w^2)]$$

$$+ p[2(\alpha u + \frac{1}{2}\beta v) - (\alpha u^2 + \frac{1}{2}\beta v^2)]$$

$$v' = 2pq[1 - 2(\alpha u + \beta v + \gamma w) + (\alpha u^2 + \beta v^2 + \gamma w^2)] + q[2(\alpha u$$

$$+ \frac{1}{2}\beta v) - (\alpha u^2 + \frac{1}{2}\beta v^2)] + p[2(\gamma w + \frac{1}{2}\beta v) - (\gamma w^2 + \frac{1}{2}\beta v^2)]$$

$$w' = q^2[1 - 2(\alpha u + \beta v + \gamma w) + (\alpha u^2 + \beta v^2 + \gamma w^2)]$$

$$+ q[2(\gamma w + \frac{1}{2}\beta v) - (\gamma w^2 + \frac{1}{2}\beta v^2)]$$

As in the model when $n=1$, at equilibrium when $n=2$, $u_*=p_*^2$, $v_*=2p_*q_*$ and $w_*=q_*^2$. After considerable algebra, the equilibrium gene frequency may be found as the solution of the cubic equation $p_*^3(\alpha+4\beta+\gamma)-p_*^2(6\beta+3\gamma)-p_*(2\alpha-6\beta-\gamma)+\gamma-2\beta=0$. In this model, if selection favours only the AA genotype, then AA has a selective coefficient of $\alpha(2-u)/[1+\alpha(1-u)(2-u)]$ and hence an advantage of $2\alpha/(1+2\alpha)$ when rare and α when common. Its selective advantage therefore becomes negatively frequency-dependent when $\alpha<\frac{1}{2}$. Generally, for larger values of n, the negative frequency-dependence becomes more and more pronounced and the equilibria become generally stable. Some special cases when $n=2$ have the following equilibria:

$$\alpha=0,\ \beta=\gamma,\ p_*=0.1806\ \text{(stable)}$$

$$\beta=0,\ \alpha=\gamma,\ p_*=\tfrac{1}{2}\ \text{(unstable)}$$

$$\gamma=0,\ \alpha=\beta,\ p_*=0.8194\ \text{(stable)}$$

$$\alpha=\beta=\gamma,\ p_*=\tfrac{1}{2}\ \text{(stable)}$$

Thus when no heterozygotes are preferred, the equilibrium is still unstable when $n=2$. However, when $\alpha=\beta=\gamma$, the equilibrium, which was neutrally stable for $n=1$, is now positively stable for $n=2$.

We should expect that as n increases, the results of sexual selection in the 'P' model should converge on those of the 'C' model with complete preferences $(n\to\infty)$. This can be tested by computing the rates of approach to equilibrium using the general probabilities that females choose to mate with the different genotypes of males. Some results are shown in table 5.5. They show that, provided there is some preference for heterozygotes, the equilibria for $n>5$ are very close to those of the model with complete preferences. This 'C' model should therefore be a realistic one when females are fairly reluctant, though not completely averse, to mate with males who are not their preferred choice. However, if there is no preference for heterozygotes, stable equilibria are not produced until $n=15$ or $n=20$. After $n=30$, the equilibria have converged closely on those of the model with complete preferences: but the females are then almost completely averse to mating except with the males they prefer; close agreement with the model with complete preferences is thus inevitable.

Table 5.5 *Genetic equilibria with particular mating preferences and different numbers of female encounters before mating*

(i) *Separate preferences for* AA *and* Aa *males but no preference for* aa *males*

Preference for AA males: $\alpha = 0.1$

No. of extra encounters n	Values of preference, β, for *Aa* males			
	0.05	0.1	0.15	0.2
1	1.0	1.0	0.750	0.667
2	1.0	0.819	0.723	0.667
3	0.920	0.786	0.713	0.667
4	0.885	0.770	0.707	0.667
5	0.886	0.762	0.704	0.667
10	0.838	0.751	0.700	0.667
15	0.834	0.750	0.700	0.667
20	0.833	0.750	0.700	0.667
∞	0.833	0.750	0.700	0.667

In this case the polymorphic equilibria are all stable: as shown in table 5.4, the stability condition for $n=1$ is $\beta > \frac{1}{2}\alpha$ which holds except when $\beta = 0.05$. When $\beta = 0.05$ and $n = 1$ or 2, fixation of A always takes place. The number of encounters, n, is the parameter in the model giving the number of extra encounters a female must make before mating with the wrong genotype of male.

(ii) *Separate preferences for* AA *and* aa *males but no preference for* Aa *males*

Preference for AA males: $\alpha = 0.1$.

No. of extra encounters n	Values of preference, γ, for *aa* males			
	0.05	0.1	0.15	0.2
1	*	*	*	*
5	*	*	*	*
10	*	0.5	*	*
15	*	0.5	0.374	*
20	0.704	0.5	0.391	0.296
30	0.674	0.5	0.399	0.326
40	0.669	0.5	0.400	0.331
50	0.667	0.5	0.400	0.333
∞	0.667	0.5	0.400	0.333

The symbol * indicates that only unstable equilibria exist for given values of n. Convergence on the equilibria of the model with complete preferences also takes much longer when $\beta = 0$. Preference for heterozygotes thus determines whether polymorphisms can be maintained in the general model with small values of n.

The values in this table are those of O'Donald (1978a).

5.2.1. *Mating preferences with different thresholds in female response*

Females may have different thresholds towards the genotypes they prefer. Then, if n extra encounters are required for random mating of females who prefer homozygotes and m extra encounters for random mating of females who prefer heterozygotes, males mate with the following frequencies:

$$P_T(AA) = \alpha[1-(1-u)^{n+1}] + \beta\left(\frac{u}{1-v}\right)(1-v)^{m+1}$$

$$+ \gamma\left(\frac{u}{1-w}\right)(1-w)^{n+1} + \delta u$$

$$P_T(Aa) = \alpha\left(\frac{v}{1-u}\right)(1-u)^{n-1} + \beta[1-(1-v)^{m+1}]$$

$$+ \gamma\left(\frac{v}{1-w}\right)(1-w)^{n+1} + \delta v$$

$$P_T(aa) = \alpha\left(\frac{w}{1-u}\right)(1-u)^{n+1} + \beta\left(\frac{w}{1-v}\right)(1-v)^{m+1}$$

$$+ \gamma[1-(1-w)^{n+1}] + \delta w$$

where $\delta = 1-\alpha-\beta-\gamma \geq 0$.

(i) *Number of encounters $n=1$ and $m=2$.* The frequencies of matings are now given by

$$P_T(AA) = u(1+\alpha-\alpha u - 2\beta v + \beta v^2 - \gamma w)$$

$$P_T(Aa) = v(1+2\beta-\alpha u - 3\beta v + \beta v^2 - \gamma w)$$

$$P_T(aa) = w(1+\gamma-\alpha u - 2\beta v + \beta v^2 - \gamma w)$$

These frequencies give the recurrence equations

$$p' = p[1-\tfrac{1}{2}(\alpha u + 2\beta v - \beta v^2 + \gamma w)] + \tfrac{1}{2}(\alpha u + \beta v - \tfrac{1}{2}\beta v^2)$$

At equilibrium, therefore

$$p = \frac{\alpha u + \beta v - \tfrac{1}{2}\beta v^2}{\alpha u + 2\beta v - \beta v^2 + \gamma w}$$

It can then be shown from the recurrence equations for u, v and w that at equilibrium when this equation holds

$$u_* = p_*^2 \qquad v_* = 2p_* q_* \qquad w_* = q_*^2$$

The genotypes are, therefore, maintained at the Hardy–Weinberg frequencies.

Put $\alpha = \gamma$ and we obtain the equation for equilibrium

$$2\beta(p-q)(1-pq) = \alpha(p-q)$$

This is a cubic equation giving three possible sets of equilibrium frequencies, as follows:

$$u_1^* = w_1^* = \tfrac{1}{4}, \quad v_1^* = \tfrac{1}{2}$$

$$u_2^* = \frac{\alpha - \beta}{2\beta} - \sqrt{\left(\frac{2\alpha - 3\beta}{4\beta}\right)}, \quad v_2^* = \frac{2\beta - \alpha}{\beta},$$

$$w_2^* = \frac{\alpha - \beta}{2\beta} + \sqrt{\left(\frac{2\alpha - 3\beta}{4\beta}\right)}$$

$$u_3^* = \frac{\alpha - \beta}{2\beta} + \sqrt{\left(\frac{2\alpha - 3\beta}{4\beta}\right)}, \quad v_3^* = \frac{2\beta - \alpha}{\beta},$$

$$w_3^* = \frac{\alpha - \beta}{2\beta} - \sqrt{\left(\frac{2\alpha - 3\beta}{4\beta}\right)}$$

Assuming that near equilibrium $u = p^2$, $v = 2pq$ and $w = q^2$, we can then differentiate the expression for Δp to obtain the conditions for the local stability of these equilibria. For the equilibrium

$$(u_1^*, v_1^*, w_1^*)$$

we obtain the condition for stability

$$\beta > \tfrac{2}{3}\alpha$$

For the equilibria (u_2^*, v_2^*, w_2^*) and (u_3^*, v_3^*, w_3^*) the condition for stability is

$$(3\beta - 2\alpha)(2\beta - \alpha) < 0$$

or since $2\beta > \alpha$ is a condition for the existence of the equilibria $\therefore$

$$3\beta/2 < \alpha < 2\beta$$

The polymorphisms exist and are stable if these conditions are met.

(ii) *Number of encounters $n = 2$ and $m = 1$.* The frequencies of matings are now given by

$$P_T(AA) = u(1 + 2\alpha - 3\alpha u + \alpha u^2 - \beta v - 2\gamma w + \gamma w^2)$$

$$P_T(Aa) = v(1 + \beta - 2\alpha u + \alpha u^2 - \beta v - 2\gamma w + \gamma w^2)$$

$$P_T(aa) = w(1 + 2\gamma - 2\alpha u + \alpha u^2 - \beta v - 3\gamma w + \gamma w^2)$$

giving the recurrence equation

$$p' = p[1 - \tfrac{1}{2}(2\alpha u - \alpha u^2 + \beta v + 2\gamma w - \gamma w^2)] + \tfrac{1}{2}(2\alpha u - \alpha u^2 + \tfrac{1}{2}\beta v)$$

Again at equilibrium the genotypes are maintained at the Hardy–Weinberg frequencies. Put $\alpha = \gamma$ and we obtain the equation for equilibrium

$$\alpha(p - q)(1 - pq) = (p - q)(2\alpha - \beta)$$

This is also cubic giving three possible sets of equilibrium frequencies, as follows:

$$u_1^* = w_1^* = \tfrac{1}{4}, \; v_1^* = \tfrac{1}{2}$$

$$u_2^* = \frac{3\alpha - 2\beta}{2\alpha} - \sqrt{\left(\frac{5\alpha - 4\beta}{4\alpha}\right)}, \quad v_2^* = \frac{2(\beta - \alpha)}{\alpha},$$

$$w_2^* = \frac{3\alpha - 2\beta}{2\alpha} + \sqrt{\left(\frac{5\alpha - 4\beta}{4\alpha}\right)}$$

$$u_3^* = \frac{3\alpha - 2\beta}{2\alpha} + \sqrt{\left(\frac{5\alpha - 4\beta}{4\alpha}\right)}, \quad v_3^* = \frac{2(\beta - \alpha)}{\alpha},$$

$$w_3^* = \frac{3\alpha - 2\beta}{2\alpha} - \sqrt{\left(\frac{5\alpha - 4\beta}{4\alpha}\right)}$$

For the equilibrium (u_1^*, v_1^*, w_1^*) we obtain the condition for stability $\beta > (5/4)\alpha$. There must, therefore, be some heterosis in the mating preference for this equilibrium to be stable. For the equilibria (u_2^*, v_2^*, w_2^*) and (u_3^*, v_3^*, w_3^*) we obtain for stability

$$(4\beta - 5\alpha)(\beta - \alpha) < 0$$

and hence for both existence and stability of these equilibria

$$\beta > \alpha > 4\beta/5$$

(iii) *Number of encounters $n = 1$ and $m \to \infty$.* The preference for heterozygotes is now complete. We have the following frequencies of matings

$$P_T(AA) = u(1 - au - \beta - \gamma w) + \alpha u$$

$$P_T(Aa) = v(1 - \alpha u - \beta - \gamma w) + \beta$$

$$P_T(aa) = w(1 - \alpha u - \beta - \gamma w) + \gamma w$$

giving the difference equation

$$\Delta p = \tfrac{1}{2}[\alpha u + \tfrac{1}{2}\beta - p(\alpha u + \beta + \gamma w)]$$

As always, the genotypes are maintained at Hardy–Weinberg frequencies at equilibrium. Then, if $\alpha = \gamma$

$$(q - p)(\tfrac{1}{2}\beta - \alpha pq) = 0$$

giving three equilibria

$$u_1^* = w_1^* = \tfrac{1}{4}, \; v_1^* = \tfrac{1}{2}$$

$$u_2^* = \frac{\alpha - \beta}{2\alpha} - \sqrt{\left(\frac{\alpha - 2\beta}{4\alpha}\right)}, \; v_2^* = \frac{\beta}{\alpha}, \; w_2^* = \frac{\alpha - \beta}{2\alpha} + \sqrt{\left(\frac{\alpha - 2\beta}{4\alpha}\right)}$$

$$u_3^* = \frac{\alpha - \beta}{2\alpha} + \sqrt{\left(\frac{\alpha - 2\beta}{4\alpha}\right)}, \; v_3^* = \frac{\beta}{\alpha}, \; w_3^* = \frac{\alpha - \beta}{2\alpha} - \sqrt{\left(\frac{\alpha - 2\beta}{4\alpha}\right)}$$

Near equilibrium

$$d\Delta p / dp = \tfrac{1}{2}(-\alpha - \beta + 3\alpha v)$$

The equilibrium (u_1^*, v_1^*, w_1^*) exists and is stable if $\alpha > \beta > \tfrac{1}{2}\alpha$. The equilibria (u_2^*, v_2^*, w_2^*) and (u_3^*, v_3^*, w_3^*) exist and are stable if $\alpha > 2\beta$.

(iv) *Number of encounters $n \to \infty$ and $m = 1$.* We have the frequencies of matings

$$P_T(AA) = u(1 - \alpha - \beta v - \gamma) + \alpha$$

$$P_T(Aa) = v(1 - \alpha - \beta v - \gamma) + \beta v$$

$$P_T(aa) = w(1 - \alpha - \beta v - \gamma) + \gamma$$

giving

$$\Delta p = \tfrac{1}{2}[\alpha + \tfrac{1}{2}\beta v - p(\alpha + \beta v + \gamma)]$$

At equilibrium, with Hardy–Weinberg frequencies, we have when $\alpha = \gamma$

$$(q - p)(\alpha + \tfrac{1}{2}\beta v) = 0$$

There is now only one equilibrium

$$u_1^* = \tfrac{1}{4}, \; v_1^* = \tfrac{1}{2}, \; w_1^* = \tfrac{1}{4}$$

The other two equilibria, which could exist, and were stable when $n = 2$ and $n = 1$ have now disappeared. The equilibrium (u_1^*, v_1^*, w_1^*) is always stable for the case when $n \to \infty$ and $m = 1$. We may conjecture that this equilibrium is globally stable.

5.3.   Natural selection of dominant and recessive characters in males

When the genotype *aa* determines a recessive character that is preferred by a proportion γ of the females and also has a selective disadvantage t as a result of increased predation, then the genotypes *AA*, *Aa* and *aa* occur with frequencies $u/(1-tw)$, $v/(1-tw)$ and $w(1-t)/(1-tw)$ among the males who may be chosen as mates. The frequencies of matings with these males are then

$$\left.\begin{array}{l} P_{\mathrm{T}}(AA)=\gamma\left(\dfrac{u}{1-w}\right)\left[1-\dfrac{w(1-t)}{1-tw}\right]^{n+1}+\dfrac{u(1-\gamma)}{1-tw} \\[3mm] P_{\mathrm{T}}(Aa)=\gamma\left(\dfrac{v}{1-w}\right)\left[1-\dfrac{w(1-t)}{1-tw}\right]^{n+1}+\dfrac{v(1-\gamma)}{1-tw} \\[3mm] P_{\mathrm{T}}(aa)=\gamma\left\{1-\left[1-\dfrac{w(1-t)}{1-tw}\right]^{n+1}\right\}+\dfrac{w(1-\gamma)}{1-tw} \end{array}\right\} \quad (5.3.1)$$

When *A* is preferred as a dominant by a proportion α of the females, and has selective disadvantage s, then

$$u(1-s)/[1-s(1-w)]$$
$$v(1-s)/[1-s(1-w)]$$

and $w/[1-s(-w)]$ are the frequencies of the genotypes in the males and

$$\left.\begin{array}{l} P_{\mathrm{T}}(AA)=\alpha\left(\dfrac{u}{1-w}\right)\left\{1-\left[\dfrac{w}{1-s(1-w)}\right]^{n+1}\right\} \\[3mm] \qquad+\dfrac{u(1-\alpha)(1-s)}{1-s(1-w)} \\[3mm] P_{\mathrm{T}}(Aa)=\alpha\left(\dfrac{v}{1-w}\right)\left\{1-\left[\dfrac{w}{1-s(1-w)}\right]^{n+1}\right\} \\[3mm] \qquad+\dfrac{v(1-\alpha)(1-s)}{1-s(1-w)} \\[3mm] P_{\mathrm{T}}(aa)=\alpha\left[\dfrac{w}{1-s(1-w)}\right]^{n+1}+\dfrac{w(1-\alpha)}{1-s(1-w)} \end{array}\right\} \quad (5.3.2)$$

When the natural selection acts only on the males and the character is not expressed in the females, the genotypes *AA*, *Aa* and *aa* are carried by the females with frequencies u, v and w. The frequencies of the matings between genotypes can then easily be calculated and hence the frequencies of the genotypes in the next generation. Mathematical expressions for the genotypic and gene frequencies can be obtained in the cases when $n=1$ and $n=2$ (O'Donald, 1978c).

Females mate at random after one extra encounter

When $n=1$, and the preferred character is recessive we obtain from equations 5.3.1 the following frequencies of the matings with the males

$$P_T(AA)=u[1-\gamma w-tw(1-\gamma)]/(1-tw)^2$$

$$P_T(Aa)=v[1-\gamma w-tw(1-\gamma)]/(1-tw)^2$$

$$P_T(aa)=w(1-t)[1-\gamma w-tw(1-\gamma)]/(1-tw)^2$$
$$+\gamma w(1-t)/(1-tw)$$

The change in gene frequency from one generation to the next can then be shown to be

$$\Delta p=\tfrac{1}{2}pw[t-\gamma(1-t)-t^2w]/(1-tw)^2 \tag{5.3.3}$$

with an equilibrium at

$$w_*=[t-\gamma(1-t)]/t^2$$

This is always unstable, however, since

$$\mathrm{d}\Delta p/\mathrm{d}p\big|_{p=p_*}=t^2p_*\, q_*\, w_*/(1-tw_*)^2$$

which must always be positive according to the premises of the model $(t>0)$. In fact the preferred character always increases in frequency if $\gamma>t/(1-t)$. No polymorphic population can be maintained in a stable state.

When the preferred character is dominant, then from equations 5.3.2 the frequencies of the matings become

$$P_T(AA)=u(1-s)\,[1+\alpha w-s(1-w)]/[1-s(1-w)]^2$$

$$P_T(Aa)=v(1-s)\,[1+\alpha w-s(1-w)]/[1-s(1-w)]^2$$

$$P_T(aa)=w[1+\alpha w-s(1-w)]/[1-s(1-w)]^2$$
$$-\alpha w/[1-s(1-w)]$$

giving a change in gene frequency of

$$\Delta p=-\tfrac{1}{2}pw[s-\alpha(1-s)-s^2(1-w)]/[1-s(1-w)]^2 \tag{5.3.4}$$

and an unstable equilibrium point

$$1 - w_* = [s - \alpha(1-s)]/s^2$$

The preferred character always increases in frequency if $\alpha > s/(1-s)$.

The difference equation for the recessive (5.3.3) can be expressed in the form

$$\Delta p = -\tfrac{1}{2} t^2 p w(w - w_*)/(1 - tw)^2$$

For the dominant, equation 5.3.4 takes the form

$$\Delta p = -\tfrac{1}{2} s^2 p w(w - w_*)/[1 - s(1 - w)]^2$$

Given the same preference for both the dominant and recessive characters and the same natural selection operating against them, the rates of selection are approximately the same. In all cases, the mating preference is either sufficient to outweigh the selective disadvantage of the preferred character which then spreads through the whole population; or it does not outweigh the selective disadvantage, and the character is eliminated.

Females mate at random after two extra encounters

When $n = 2$, we derive from equations 5.3.1 the following frequencies of matings when the preference favours a recessive character

$$P_T(AA) = u[1 - 2w(\gamma + t - \gamma t) + w^2(\gamma + t^2 - \gamma t^2)]/(1 - tw)^3$$

$$P_T(Aa) = v[1 - 2w(\gamma + t - \gamma t) + w^2(\gamma + t^2 - \gamma t^2)]/(1 - tw)^3$$

$$\begin{aligned} P_T(aa) = \ & w[1 - 2w(\gamma + t - \gamma t) + w^2(\gamma + t^2 - \gamma t^2)]/(1 - tw)^3 \\ & + \gamma w(1 - w)(1 - t)/(1 - tw)^3 + \gamma w(1 - t)/(1 - tw)^2 \\ & - tw/(1 - tw) \end{aligned}$$

giving the change in gene frequency

$$\Delta p = \tfrac{1}{2} p w[-2\gamma + t + 2\gamma t + w(\gamma - 2t^2 - \gamma t^2) \tag{5.3.5}$$
$$+ t^3 w^2]/(1 - tw)^3$$

The equilibrium frequency of the preferred phenotypes is then given by the solution of the quadratic equation

$$t^3 w^2 + w(\gamma - 2t^2 - \gamma t^2) - 2\gamma + t + 2\gamma t = 0$$

Since the first term is usually negligible, a good approximation to the equilibrium frequency is given by

$$w_* = (2\gamma - t - 2\gamma t)/(\gamma - 2t^2 - \gamma t^2) \qquad (5.3.6)$$

Using this approximation in equation 5.3.5 we get

$$\Delta p = \tfrac{1}{2} pw(w - w_*)(\gamma - 2t^2 - \gamma t^2)/(1 - tw)^3$$

At equilibrium $w_* = q_*^2$ and since $dw/dp = -2q$ near equilibrium, therefore after differentiation

$$d\Delta p/dp|_{p=p_*} = -(\gamma - 2t^2 - \gamma t^2)p_* q_* w_*/1 - tw_*)^3$$

showing that the equilibrium will be stable if $\gamma - 2t^2 - \gamma t^2 > 0$ or $\gamma > 2t^2/(1 - t^2)$. If $t = 0.05$, stable polymorphic equilibrium values of w_* are found, as shown in table 5.6. Equation 5.3.6 is thus fairly satisfactory at these values of γ and t. Polymorphic equilibria exist within the range of values

$$\tfrac{1}{2} t/(1 - t) < \gamma < t(1 - 2t)/(1 - t)^2$$

The preferred character increases in frequency if $\gamma > \tfrac{1}{2} t/(1 - t)$. This condition also implies stability of the polymorphic equilibrium if $t < \tfrac{1}{3}$.

When a dominant character is preferred then from equations 5.3.2 we derive the following frequencies of matings

$$P_T(AA) = u(1 - s)[(1 - s)^2 + w(1 - s)(\alpha + 2s)$$
$$+ w^2(\alpha + \alpha s + s^2)]/[1 - s(1 - w)]^3$$

Table 5.6 *Equilibrium frequencies of recessive according to exact and approximate solutions*

Values of γ	Values of w_*	
	Exact	By approximte equation 5.3.6
0.025	0	0
0.030	0.2804	0.2808
0.035	0.5503	0.5516
0.040	0.7430	0.7450
0.045	0.8875	0.8900
0.050	1.0	1.0

$$P_T(Aa) = v(1-s)[(1-s)^2 + w(1-s)(\alpha+2s)$$
$$+ w^2(\alpha+\alpha s+s^2)]/[1-s(1-w)]^3$$
$$P_T(aa) = w[(1-s)^2 + w(1-s)(\alpha+2s)$$
$$+ w^2(\alpha+\alpha s+s^2)]/[1-s(1-w)]^3$$
$$- \alpha w[(1-s)^2 + w(1-s)(1+2s)$$
$$+ sw^2(1+s)]/[1-s(1-w)]^3$$

hence the equation for the change of gene frequency

$$\Delta p = -\tfrac{1}{2}pw[-2\alpha+s+2\alpha s+(1-w)(\alpha-2s^2-\alpha s^2) \tag{5.3.7}$$
$$+ s^3(1-w)^2]/[1-s(1-w)]^3$$

and a similar equilibrium frequency of the preferred phenotype, under the same conditions, as for a recessive character. Approximately therefore

$$1-w_* = (2\alpha-s-2\alpha s)/(\alpha-2s^2-\alpha s^2) \tag{5.3.8}$$

which will have the same error for given values of α and s as equation 5.3.6 for the corresponding parameters of the recessive character.

Females mate at random after a number of extra encounters

If the females mate preferentially at a much lower threshold than they mate at random, extra encounters will be needed to raise their level of stimulation from the preferential to the random mating threshold. As the number of these extra encounters increases, so does the advantage gained by the rare males: the sexual selection becomes more strongly frequency-dependent in a negative sense, and polymorphic equilibria become stable over a wider range of values. As $n \to \infty$, the females with preferences always mate preferentially giving the 'C' models already analysed in Chapter 4. Then the equilibrium frequencies are

$$w_* = \gamma/t$$

for the recessive character and

$$1-w_* = \alpha/s$$

for the dominant character (equations 4.3.5 and 4.3.6).

In general, if a selectively deleterious character is the preferred phenotype, then for any value of n its frequency at equilibrium, w_*, can be found as a solution of the equation

$$\gamma[(1-w)/(1-tw)]^n = \gamma - tw \tag{5.3.9}$$

If the character is dominant, its equilibrium frequency, $1-w_*$, can be found from the equation

$$\alpha[w/(1-s+sw)]^n = \alpha - s(1-w) \tag{5.3.10}$$

These equations reduce to those already found for the equilibria when $n=1$ or $n=2$. In all the models analysed, when sexual selection is balanced by natural selection, the phenotypic equilibrium frequencies are the same for both dominant and recessive characters. Equilibria for particular numerical values of mating preference and natural selection can easily be computed for any value of n. Table 5.7 shows the results of such computations. The equilibria given by equations 5.3.9 and 5.3.10 are stable; they rapidly approach the corresponding equilibria established with complete preferences. When $n>5$, the equilibrium frequency is very close to that determined by complete preference. This suggests that the many models of sexual selection and assortative mating based on the concept of complete preferences should give widely applicable results.

Since the females will vary in the expression of their preference towards different males, we should expect that n would vary between females with preferences for different males. Such variation will give rise to a model in which the frequencies of the matings are given by

$$P_T(A) = \alpha(1 - y^{n+1}) + \gamma x^{m+1} + x(1 - \alpha - \gamma)$$

$$P_T(B) = \alpha y^{n+1} + \gamma(1 - x^{m+1}) + y(1 - a - \gamma)$$

Particular cases of this general model have already been analysed in sections 5.1.1 and 5.2.1. These were cases in which one of the encounter parameters (n or m) took the value 1 and the other parameter took the value 2 or was increased without limit to give rise to complete preference for either one of the preferred phenotypes or the other. Separate preferences for each of the genotypes were shown to give rise to three different stable equilibria.

In the most general model, some females prefer the dominant and others the recessive character, while natural selection acts in

Table 5.7 *Phenotypic equilibrium frequencies of a character sex-limited to males and maintained by female preference and selective disadvantage*

No. of extra encounters n	Mating preferences for a character with a selective disadvantage of 10 per cent								
	0.01	0.02	0.03	0.04	0.05	0.06	0.07	0.08	0.09
1	0	0	0	0	0	0	0	0	0
2	0	0	0	0	0	0.202	0.522	0.734	0.886
3	0	0	0	0.094	0.342	0.524	0.669	0.791	0.899
4	0	0	0.061	0.279	0.442	0.575	0.692	0.798	0.900
5	0	0	0.174	0.341	0.474	0.591	0.697	0.800	0.900
10	0	0.158	0.286	0.396	0.499	0.600	0.700	0.800	0.900
20	0.076	0.196	0.300	0.400	0.500	0.600	0.700	0.800	0.900
∞	0.1	0.2	0.3	0.4	0.5	0.6	0.7	0.8	0.9

The polymorphic equilibrium frequencies shown in this table are all globally stable, but are reached only slowly at the values when $n=1$ and $n=2$.

favour of either one or other character. The equilibrium frequency can then be found by solving the equation

$$\alpha w(1-t)\left\{1-\left[\frac{w(1-t)}{1-tw}\right]^n\right\}$$

$$-\gamma(1-w)\left[1-\left(\frac{1-w}{1-tw}\right)^m\right]+tw(1-w)=0$$

In this equation, natural selection favours the dominant if $t>0$ and the recessive if $t<0$. If $t<0$, the dominant is then at a selective disadvantage given by $s=-t/(1-t)$. Putting $\alpha=0$ gives equation 5.3.9 and putting $\gamma=0$ gives equation 5.3.10. In the case when $m=n=1$, there is an unstable equilibrium at the point

$$w_*=[t+(\alpha+\gamma)(1-t)]/t^2$$

The change in gene frequency is given by

$$\Delta p=\tfrac{1}{2}pw[t+(\alpha+\gamma)(1-t)-t^2w]/(1-tw)^2$$

$$=-\tfrac{1}{2}t^2pw(w-w_*)/(1-tw)^2$$

showing that aa fixes if $w>w_*$ and AA fixes if $w<w_*$. When $m=n=2$, the equation for equilibrium of aa becomes

$$t^3w^2+w\{(1-t)[\alpha(1-2t)+\gamma(1+t)]-2t^2\}$$

$$+(1-t)(\alpha-2\gamma)+t=0$$

This equation gives the equilibria already found (equations 5.3.5 and 5.3.6, 5.3.7 and 5.3.8) when either $\alpha=0$ or $\gamma=0$.

5.4. Natural selection of dominant and recessive characters in both sexes

When the preferred character is expressed in both sexes, the females will also be subject to natural selection. They mate, after the natural selection has taken place, with the same population frequencies of the genotypes as the males. However, the frequencies at which the females as a whole choose to mate with males who are AA, Aa or aa are just the same as those when natural selection acts only on males: these overall mating frequencies are determined by the same probabilities of encountering the different males. Thus the equations for $P_T(AA)$, $P_T(Aa)$ and $P_T(aa)$ are the same as in the corresponding models analysed in section 5.3 (equations 5.3.1 and

5.3.2). Mathematical solutions can only be obtained when $n=1$ or $n=2$ (O'Donald, 1978c).

Females mate at random after one extra encounter

The change in gene frequency when $n=1$ and females prefer a recessive character is given by

$$\Delta p = pw[t - \tfrac{1}{2}\gamma(1-t) - t^2 w]/(1-tw)^2 \tag{5.4.1}$$

with an unstable equilibrium of the preferred phenotypes

$$w_* = [t - \tfrac{1}{2}\gamma(1-t)]/t^2$$

If the mating preference were doubled, this would be the same expression as when selection acts only on males. The preferred character only spreads through the population if $\gamma > 2t/(1-t)$: as we should expect, therefore, the sexual selection must be twice as great to outweigh the effects of natural selection that acts on both sexes and not just on males.

When the females prefer a dominant character

$$\Delta p = -pw[s - \tfrac{1}{2}\alpha(1-s) - s^2(1-w)]/[1-s(1-w)]^2$$

$$1 - w_* = [s - \tfrac{1}{2}\alpha(1-s)]/s^2$$

and the preferred character only increases if $\alpha > 2s/(1-s)$.

Females mate at random after two extra encounters

When $n=2$ and females prefer a recessive, a great deal of algebra is required to show that

$$\Delta p = pw[-\gamma + t + \gamma t + w(\tfrac{1}{2}\gamma - 2t^2 - \tfrac{1}{2}\gamma t^2) + t^3 w^2]/(1-tw)^3 \tag{5.4.2}$$

This equation can be derived intuitively from equation 5.3.5: the overall rate of selection is twice as great as when natural selection acts only on males and the effect of the mating preference is halved. The condition for initial increase of the preferred character is $\gamma > t/(1-t)$ compared to the condition $\gamma > \tfrac{1}{2}t(1-t)$ when natural selection acts only on males. The equilibrium, as expected, is now stable if $\gamma > 4t^2/(1-t^2)$, and is reached at an approximate frequency

$$w_* = (\gamma - t - \gamma t)/(\tfrac{1}{2}\gamma - 2t^2 - \tfrac{1}{2}\gamma t^2) \tag{5.4.3}$$

which may be compared with equation 5.3.6 for natural selection acting only on males. The condition for initial increase implies the stability condition for all values of $t < \frac{1}{3}$.

When the females prefer a dominant character, we have the corresponding equations

$$\Delta p = -pw[-\alpha + s + \alpha s + (1-w)(\tfrac{1}{2}\alpha - 2s^2 - \tfrac{1}{2}\alpha s)$$
$$+ s^3(1-w)^2]/[1 - s(1-w)]^3 \qquad (5.4.4)$$

$$1 - w_* = (\alpha - s - \alpha s)/(\tfrac{1}{2}\alpha - 2s^2 - \tfrac{1}{2}\alpha s^2) \qquad (5.4.5)$$

with conditions for initial increase and stability $\alpha > s/(1-s)$ and $\alpha > 4s^2/(1-s^2)$ implying stability for values $s < \frac{1}{3}$. These results may again be compared with those given by the corresponding equations 5.3.7 and 5.3.8 for natural selection acting only on males.

5.5   Fitting the models to data

The 'P' models I have analysed can be fitted to the data of frequency-dependent matings of male *Drosophila* (Ehrman, 1967, 1970; Spiess, 1968; Spiess and Spiess, 1969). In fitting the general model to the same data, the additional parameter, n, must therefore be estimated as well as the parameters α and γ of the preferences for the two genotypes of males of which females were offered a choice. Given two males, A and B, at frequencies u and v fixed in each experiment, the probabilities that females then choose to mate with them are as follows:

Male genotypes	AA	BB
Frequencies	u	v
Preferences	α	γ
Probabilities that females choose to mate with these males	$\alpha(1 - v^{n+1}) + \gamma u^{n+1}$ $+ u(1 - \alpha - \gamma)$	$\alpha v^{n+1} + \gamma(1 - u^{n+1})$ $+ v(1 - \alpha - \gamma)$

As in the fitting of the data to the model with complete preferences, the maximum likelihood estimates of α and γ were found by trial and error for a range of values of n. Tables 5.8 and 5.9 give the results of fitting the model to Ehrman's and Spiess' data on *Drosophila*. In Spiess' data on the mating advantage for the homozygous karyotypes AR and PP, the maximum log likelihood occurs when $n = 3$ with preferences for AR and PP of $\alpha = 0.418$ and $\gamma = 0.192$. We cannot estimate any preference that may exist for AR/PP heterozygotes; but in the absence of any such preference no

Table 5.8 *Fitting data of Spiess and Spiess to the model of mating behaviour*

(i) *Matings with karyotypes homozygous for inversions* AR *and* PP *in* Drosophila pseudoobscura (*Spiess*, 1968)

No. of extra encounters n	Estimates of preferences at maximum likelihood AR	PP	log likelihood	Residual χ^2
0	–	–	–	67.105
1	0.359	0.028	-805.532	33.132
2	0.634	0.366	-793.848	6.272
3	0.418	0.192	-793.584	5.620
4	0.316	0.115	-793.594	5.637
5	0.267	0.082	-793.612	5.671
10	0.185	0.037	-793.944	6.323
20	0.151	0.025	-794.638	7.674
30	0.142	0.022	-794.961	8.301
∞	0.138	0.021	-795.151	8.672

The maximum likelihood is attained at $n=3$, but the increase in log likelihood over that for complete preferences ($n\to\infty$) is only 1.567 which is not significant. The residual χ^2 has 6 degrees of freedom and measures the heterogeneity remaining in the data after the model has been fitted.

(ii) *Matings with strains* Humboldt (*Hu*) *and* White Wolf (*WW*) *in* Drosophila persimilis (*Spiess and Spiess*, 1969)

No. of extra encounters n	Estimates of preferences at maximum likelihood Hu	WW	log likelihood	Residual χ^2
0	–	–	–	101.707
1	0.220	0.172	-492.207	101.053
3	0.508	0.492	-458.278	15.523
5	0.448	0.440	-453.946	6.140
7	0.335	0.326	-454.294	6.873
10	0.262	0.252	-454.918	8.185
20	0.190	0.180	-456.467	11.497
30	0.173	0.164	-457.109	12.900
40	0.168	0.158	-457.342	13.414
∞	0.166	0.156	-457.467	13.692

The maximum likelihood is attained at $n=5$ representing a highly significant increase in log likelihood of 3.521 over that for complete preferences. This shows that fitting the general model significantly reduces the residual heterogeneity in the data.

O'Donald (1978a) gave the values shown in this table.

Table 5.9   *Fitting data of Ehrman on matings of* Drosophila pseudoobscura *to the model of mating behaviour*

(i) *Matings of karyotypes homozygous for inversions* CH *and* AR *in a line that had been selected for negative geotaxis in previous generations* (Ehrman, 1967)

No. of extra encounters	Estimates of preferences at maximum likelihood		log likelihood	Residual χ^2
n	CH	AR		
0	–	–	–	170.080
1	0.158	0.261	− 729.715	170.540
3	0.458	0.542	− 682.989	46.166
5	0.462	0.538	− 671.179	19.022
6	0.450	0.523	− 670.578	17.537
7	0.404	0.473	− 670.584	17.522
10	0.332	0.394	− 670.782	17.881
20	0.266	0.318	− 671.852	20.012
30	0.250	0.300	− 672.471	21.258
∞	0.243	0.292	− 672.860	22.044

The maximum likelihood occurs at $n=6$, but minimum χ^2 at $n=7$. The log likelihood at $n=6$ represents a significant increase of 2.282 over that for complete preferences.

(ii) *Matings of karyotypes homozygous for inversions* CH *and* AR *in a line that had been selected for positive geotaxis in previous generations* (Ehrman, 1967)

No. of extra encounters	Estimates of preferences at maximum likelihood		log likelihood	Residual χ^2
n	CH	AR		
0	–	–	–	244.051
1	0.021	0.232	− 721.740	211.856
5	0.446	0.554	− 659.927	39.063
10	0.316	0.422	− 657.570	32.033
15	0.264	0.372	− 657.099	30.724
20	0.242	0.352	− 656.993	30.376
25	0.231	0.341	− 656.996	30.323
30	0.225	0.335	− 657.023	30.348
40	0.220	0.330	− 657.065	30.411
∞	0.217	0.328	− 657.095	30.460

The maximum likelihood occurs at $n=20$ but the fit is no significant improvement over that for complete preferences.

Table 5.9   *continued*

(iii) *Matings of lines homozygous for the recessive alleles* or *(orange) and* pr *(purple)* (*Ehrman*, 1970)

No. of extra encounters	Estimates of preferences at maximum likelihood		log likelihood	Residual χ^2
n	*or*	*pr*		
0	–	–	–	64.132
1	0.222	0.229	– 867.152	64.111
5	0.369	0.383	– 853.019	31.055
10	0.267	0.282	– 852.855	30.743
15	0.250	0.265	– 852.852	30.756
20	0.246	0.261	– 852.867	30.793
25	0.244	0.259	– 852.874	30.812
30	0.244	0.259	– 852.877	30.819
40	0.244	0.259	– 852.879	30.822
∞	0.244	0.259	– 852.879	30.823

In data on matings of *or* and *pr*, there were two lines started at 80 *or*: 20 *pr* and 20 *or*: 80 *pr* and continued for 10 generations in which the frequencies of matings in one generation determined the frequencies of individuals in the next, both lines reaching about 50 per cent *or*. The residual χ^2 has 17 degrees of freedom.

The values in this table are also those of O'Donald (1978a).

polymorphic equilibrium would be stable, as table 5.5 (ii) shows. Stable equilibria would be produced, however, if preference for one of the karyotypes were dominant over the other. The model with dominance then predicts that a stable polymorphism would be attained at $p_* = 0.609$.

In Ehrman's data on sexual selection for different karyotypes of *Drosophila pseudoobscura* significant heterogeneity, which had been revealed by the previous analyses of the data (O'Donald, 1977a, 1977c), remains after fitting the more general model. This was to be expected since the deviations from the expectations of the model with complete preferences were not consistent: mating at one or two intermediate frequencies showed a great excess of the rarer males' mating advantage. I suggested that if the preferences were genetically determined, they may fluctuate by chance between one set of matings and another thus raising the heterogeneity above the level set by constant mating preferences.

The predicted equilibrium gene frequency for the preferences estimated for Ehrman's data is given approximately by the equations for complete preferences:

(i) for the model with separate preferences for each genotype and no dominance

$$p_* = (\alpha + \tfrac{1}{2}\beta)/(\alpha + \beta + \gamma);$$

(ii) for the model with dominance of the allele A

$$p_* = 1 - \sqrt{[\gamma/(\alpha + \gamma)]}$$

Assuming that the model without dominance can be applied, we find for the equilibrium frequency of CH in the negatively selected line (table 5.9(i))

$$p_* = 0.454$$

and for the positively selected line (table 5.9(ii))

$$p_* = 0.398$$

According to the model with dominance, the corresponding frequencies would be

$$p_* = 0.261 \text{ (negatively selected line)}$$

$$p_* = 0.224 \text{ (positively selected line)}$$

In population cage experiments carried out over many generations, the equilibrium gene frequency attained and the rate of approach to equilibrium would provide independent tests of the model and independent estimates of the preferences. The equilibria predicted by the models of sexual selection would of course be disturbed by any natural selection of the genotypes; and unfortunately, independent estimates of the coefficients of natural selection would be almost impossible to obtain. In this respect, since sexual selection arises because individuals differ only in their chances of mating – a simpler process, more easily observed than the various possible components of natural selection – the theory of sexual selection is more easily refuted than the theory of natural selection. Once the theoretical basis of sexual selection is established therefore, the details of its mechanisms should more rapidly be determined and its role in evolution more completely understood.

6

Frequency-dependent expression of mating preference

6.1. Positive frequency-dependent preferences for dominant and recessive characters

In analysing the effects of partial mating preferences, we assumed that if a female encountered at least one male she preferred in $n+1$ males she would mate preferentially with him. In the mathematical derivation of the frequencies of the matings with the males, it is irrelevant whether the $n+1$ males are encountered successively or in groups. In the models we shall now analyse, the expression of preference is assumed to be proportional to the number of preferred males encountered in the group of $n+1$ males, either with positive frequency-dependence if a preference is exercised more often as more of the preferred males are encountered, or with negative frequency-dependence if the preference is exercised less often with more encounters. Encounters with the preferred males are assumed either to facilitate or to inhibit the expression of preference. As shown in Chapter 3, sections 3.3 and 3.4, in the limiting case as $n \to \infty$, these frequency-dependent, or 'PF', models can be derived directly from the 'C' models with complete preference by assuming either that the preferences are proportional to the frequency of the preferred males giving the positive frequency-dependent, or 'PF+' models, or that the preferences are proportional to the frequency of all the other males who do not possess the preferred phenotypes giving the negative frequency-dependent, or 'PF−' models. In these limiting cases, the females' preferences are expressed according to their frequency of encounter with an unlimited number of males and hence are proportional to the population frequency of the preferred males in the 'PF+' models or to the population frequency of all but the preferred males in the 'PF−' models.

In section 2.4, I discussed behavioural mechanisms that might produce these frequency-dependent responses. According to Ehrman and Spiess' theory (Ehrman and Spiess, 1969; Spiess and Ehrman, 1978), the females sample 'courtship cues' from the males, become habituated to the commoner males and then accept a rarer male with a different cue because he interrupts the habituation. I have suggested (O'Donald, 1978b; and in section 2.4) that such behaviour would be mal-adaptive: it would entail the greatest female preference for the rarest, and probably therefore the most deleterious, males – males whose phenotypes were the most deviant from the normal or the fittest. I am also uncertain how far Ehrman and Spiess' theory represents a genuine explanation – that is, an hypothesis about causal factors which are independent of, and at the same time entail, the consequences to be explained – and how far it is merely a tautology, or restatement of these consequences in a new set of terms. Averhoff and Richardson's theory (1976) postulates that concentrations of pheromones determine the expression of preference for specific genotypes. And I have argued (in section 2.4) that this might produce either negative or positive frequency-dependent preferences: if concentrations of a pheromone act up to a threshold at which receptor cells become accommodated and no longer respond, a smaller proportion of females will respond at higher concentrations; if a lower threshold must be exceeded before the receptors are activated, a greater proportion will respond at higher concentrations. In more general terms, we may suppose that previous encounters with courting males determine the females' thresholds. A female might have a higher threshold towards some genotypes and therefore mate preferentially at a lower threshold with males of other genotypes. With constant thresholds, the frequencies of the matings will be given by the 'P' models. But suppose previous encounters with the preferred males raise the initially low threshold against them: the more often preferred males have been encountered, the less the difference in the thresholds ànd the lower the frequency of preferential matings. This mechanism would give rise to a frequency-dependent expression of preference similar to the 'PF −' models. Alternatively, differences between thresholds might not exist initially but arise as certain phenotypes of males are encountered. If female encounters with certain males raised a threshold against the others, matings would take place at frequencies given by the 'PF +' models. In both of these models of mating response, the males, who have previously been encountered and who have therefore determined the degree of expression of the preference, need not be those from whom the females will choose their mates. In neither of these models do the females express a general frequency-dependent preference for any phenotype of male, such as Spiess and Ehrman seem to assume will be expressed for all rare males: preferences are shown only for particular phenotypes of the males. If females did express a general, frequency-dependent preference, I should expect that it would favour the commoner, rather than, as Spiess and Ehrman suggest, the rarer phenotypes: the commoner phenotypes will more often have an overall selective advantage in their favour; and females with a preference for advantageous males will also gain an advantage for themselves by producing sons who possess the advantageous characters. This was Fisher's suggestion (Fisher, 1930) of the 'runaway process' of sexual selection in which selection for a mating preference is continually reinforced by the increasing advantage of the preferred character. Zahavi (1975), as we shall see in section 8.2, made the opposite and contradictory assumption that females should prefer males with selectively disadvantageous 'handicaps', for handicapped males will have been subjected to more intense natural selection and must therefore have been the fitter to survive. He ignored the disadvantage of such matings in producing handicapped sons. In fact, these matings must be at an overall disadvantage compared to matings with normal males. These and other models of the evolution of mating preferences will be analysed in Chapter 8.

If positive, frequency-dependent mating preferences may be exercised in favour of two male phenotypes A and B, then as shown in section 3.3 matings with these males take place at frequencies given by equations 3.3.1 which are as follows:

$$P_T(A) = x + nxy(\alpha - \gamma)/(n+1) \qquad P_T(B) = y - nxy(\alpha - \gamma)/(n+1)$$

As before, A and B occur with population frequencies x and y, and proportions α and γ of the females have preferences for them. The preferences are exercised with probabilities given by the proportion of the preferred males who were encountered in a group of $n+1$ males. If A is a dominant character and B is recessive, then we obtain frequencies of matings with male genotypes

$$\left.\begin{aligned}
P_T(AA) &= u + nuw(\alpha - \gamma)/(n+1) \\
P_T(Aa) &= v + nvw(\alpha - \gamma)/(n+1) \\
P_T(aa) &= w + nw^2(\alpha - \gamma)/(n+1) - nw(\alpha - \gamma)/(n+1)
\end{aligned}\right\} \qquad (6.1.1)$$

The frequencies of genotypes in the next generation are then given by

$$u' = p^2[1 + nw(\alpha - \gamma)/(n+1)]$$

$$v' = 2pq[1 + nw(\alpha - \gamma)/(n+1)] - npw(\alpha - \gamma)/(n+1)$$

$$w' = q^2[1 + nw(\alpha - \gamma)/n+1)] - nqw(\alpha - \gamma)/(n+1)$$

and hence the change in gene frequency from one generation to the next by

$$\Delta p = \tfrac{1}{2}npw(\alpha - \gamma)/(n+1) \qquad (6.1.2)$$

Obviously no polymorphic equilibrium can be maintained. If $\alpha > \gamma$, $p_n \to 1$ and the allele A is ultimately fixed in the population. If $\alpha < \gamma$, $p_n \to 0$ and A is eliminated. In the case in which $n \to \infty$,

$$\Delta p = \tfrac{1}{2}pw(\alpha - \gamma)$$

and the model is mathematically identical to the corresponding 'P' model in which females with preferences need to make only one extra encounter with a courting male before they mate at random. The 'PF+' model in which $n \to \infty$ is thus equivalent to the 'P' model with $n = 1$. This is because in the 'P' model with $n = 1$, the probability of preferential mating is equal to the probability of an encounter with one of the preferred males and hence equal to the population frequency of these males. This is also the probability that females mate preferentially in the 'PF+' models since the proportion of preferred phenotypes in a large group becomes equal to their population frequency as the size of the group is increased without limit.

To a good approximation we may put $w = q^2$ in equation 6.1.2 to obtain

$$\Delta p = \tfrac{1}{2}npq^2(\alpha - \gamma)/(n+1)$$

Therefore, when $\alpha - \gamma$ and hence Δp are small quantities, we can obtain the gene frequency after a number of generations by integration giving the equation

$$\tfrac{1}{2}nT(\alpha - \gamma)/(n+1) = \log_e \left(\frac{p_T q_0}{p_0 q_T} \right) + \frac{1}{q_T} - \frac{1}{q_0}$$

As in previous sections, T is the number of generations in which the gene frequency has changed from p_0 to p_T.

6.2 Positive frequency-dependent preferences for each genotype

When females may exercise preferences for particular genotypes, I assume they do not distinguish between the other genotypes. In section 3.3, I showed that among females with preferences for males with phenotype A, a proportion

$$P'_m = x$$

mate only with the males they prefer while proportions

$$P_m(A) = nxy/(n+1)$$

and

$$P_m(B) = y(ny + 1)/(n+1)$$

do not exercise their preference and mate randomly with A and B males. If α, β and γ of the females prefer AA, Aa and aa males respectively, the following frequencies of matings take place with the groups of females with these preferences.

Females with a preference for AA males

$$P'_m = u$$

$$P_m(AA) = nu(1-u)/(n+1)$$

$$P_m(Aa) = \left\{ \frac{(1-u)[n(1-u)+1]}{n+1} \right\} \cdot \left(\frac{v}{1-u} \right)$$

$$= v(n - nu + 1)/(n+1)$$

$$P_m(aa) = w(n - nu + 1)/(n+1)$$

Females with a preference for Aa males

$$P'_m = v$$

$$P_m(AA) = u(n - nv + 1)/(n+1)$$

$$P_m(Aa) = nv(1-v)/(n+1)$$

$$P_m(aa) = w(n - nv + 1)/(n+1)$$

Females with a preference for aa males

$$P'_m = w$$

$$P_m(AA) = u(n - nw + 1)/(n+1)$$

$$P_m(Aa) = v(n - nw + 1)/(n + 1)$$

$$P_m(aa) = nw(1 - w)/(n + 1)$$

Therefore the overall frequency of matings with AA males is given by the equation

$$P_T(AA) = \alpha u + [n\alpha u(1 - u) + \beta u(n - nv + 1)$$

$$+ \gamma u(n - nw + 1)]/(n + 1) + u(1 - \alpha - \beta - \gamma)$$

$$= u[1 + n(\alpha - \alpha u - \beta v - \gamma w)/(n + 1)]$$

Similarly, for Aa and aa males

$$\left. \begin{array}{l} P_T(Aa) = v[1 + n(\beta - \alpha u - \beta v - \gamma w)/(n + 1)] \\ P_T(aa) = w[1 + n(\gamma - \alpha u - \beta v - \gamma w)/(n + 1)] \end{array} \right\} \quad (6.2.1)$$

Hence, for the frequencies of the genotypes in the next generation

$$u' = p^2[1 - n(\alpha u + \beta v + \gamma w)/(n + 1)] + np(\alpha u + \tfrac{1}{2}\beta v)/(n + 1)$$

$$v' = 2pq[1 - n(\alpha u + \beta v + \gamma w)/(n + 1)] + nq(\alpha u + \tfrac{1}{2}\beta v)/(n + 1)$$

$$+ np(\gamma w + \tfrac{1}{2}\beta v)/(n + 1)$$

$$w' = q^2[1 - n(\alpha u + \beta v + \gamma w)/(n + 1)] + nq(\gamma w + \tfrac{1}{2}\beta v)/(n + 1)$$

and for the change in gene frequency from the one generation to the next

$$\Delta p = \tfrac{1}{2}n[\alpha u + \tfrac{1}{2}\beta v - p(\alpha u + \beta v + \gamma w)]/(n + 1) \quad (6.2.2)$$

Thus at equilibrium

$$p_* = (\alpha u_* + \tfrac{1}{2}\beta v_*)/(\alpha u_* + \beta v_* + \gamma w_*)$$

showing that

$$u_* = p_*^2$$

$$v_* = 2p_* q_*$$

$$w_* = q_*^2$$

Substituting these Hardy–Weinberg frequencies for u_*, v_* and w_*, we get

$$(\alpha - 2\beta + \gamma)p_*^2 - (\alpha - 3\beta + 2\gamma)p_* + \gamma - \beta = 0$$

which is the same as

$$(p_* - 1)[(\alpha - 2\beta + \gamma)p_* + \beta - \gamma] = 0$$

so that polymorphic equilibrium exists with frequencies

$$\left. \begin{array}{l} p_* = (\gamma - \beta)/(\alpha - 2\beta + \gamma) \\ q_* = (\alpha - \beta)/(\alpha - 2\beta + \gamma) \end{array} \right\} \tag{6.2.3}$$

Assuming that near equilibrium, the genotypic frequencies are very close to those of Hardy–Weinberg and substituting these frequencies in 6.2.2 then we find that approximately

$$\Delta p = \tfrac{1}{2}npq[(\alpha - \beta) - q(\alpha - 2\beta + \gamma)]/(n+1)$$

Differentiating this expression we get

$$d\Delta p/dp|_{p=p_*} = \tfrac{1}{2}np_*q_*(\alpha - 2\beta + \gamma)/(n+1)$$

showing that the polymorphic equilibrium is stable if $\alpha - 2\beta + \gamma > 0$ or $\beta > \tfrac{1}{2}(\alpha + \gamma)$. Thus for stability, more females must prefer *Aa* males than the average of those that prefer *AA* or *aa* males. This condition always holds if $\beta > \max(\alpha, \gamma)$.

The rate of approach to equilibrium can be found by putting the difference equation 6.2.2 in the form

$$\Delta p = -\tfrac{1}{2}npq(q - q_*)(\alpha - 2\beta + \gamma)/(n+1)$$

treating it as a differential equation and integrating it. This gives the following equation for the gene frequency after T generations

$$\left(\frac{q_0}{q_T}\right)^{p_*} \cdot \left(\frac{p_0}{p_T}\right)^{q_*} \cdot \left(\frac{q_T - q_*}{q_0 - q_*}\right)$$

$$= \exp[\tfrac{1}{2}nTp_*q_*(\alpha - 2\beta + \gamma)/(n+1)] \tag{6.2.4}$$

showing that $q_T \to q_*$ at a geometric rate so long as $\beta > \tfrac{1}{2}(\alpha + \gamma)$. For small values of Δp, the expression $\exp[\tfrac{1}{2}nTp_*q_*(\alpha - 2\beta + \gamma)/(n+1)]$ can be used to give the values of the left-hand side of equation 6.2.4 given any initial frequency p_0, since in this case convergence on the equilibrium value is global. Thus values of p_T observed in successive generations can be used to fit the model according to the expected values of the expression

$$\left(\frac{q_0}{q_T}\right)^{p_*} \cdot \left(\frac{p_0}{p_T}\right)^{q_*} \cdot \left(\frac{q_T - q_*}{q_0 - q_*}\right)$$

in each generation. As $n \to \infty$, equation 6.2.4 becomes identical to equation 5.2.3 in the corresponding 'P' model with $n = 1$.

This analysis shows that in the 'PF+' models, sexual selection can only maintain a stable polymorphism if a preference for heterozygotes outweighs the preference for other genotypes. In other cases, one allele must have an advantage over the other allele leading to fixation of the one and elimination of the other. In the more realistic models, when the preferred phenotype is determined by a dominant or recessive allele, no stable polymorphism ever exists.

6.3   Positive frequency-dependence and natural selection in males

In these models the males mate after natural selection, acting to the disadvantage of the preferred phenotypes, has changed their frequencies. If natural selection acts against the B phenotype determined by the recessive allele a, then the males mate with population frequencies $u/(1-tw)$, $v/(1-tw)$ and $w(1-t)/(1-tw)$ of the genotypes AA, Aa and aa, where t measures the relative selective disadvantage of the genotype aa. In section 6.1, I derived the frequencies of matings with male genotypes at population frequencies u, v and w (equations 6.1.1). These are zygotic frequencies unaltered by the effects of natural selection. To determine the frequencies of matings if selection has taken place in the males, we may substitute the frequencies after selection in place of the unaltered frequencies u, v and w. This gives the equations for the matings of the recessive

$$P_T(AA) = u[1 - tw - n\gamma w(1-t)/(n+1)]/(1-tw)^2$$
$$P_T(Aa) = v[1 - tw - n\gamma w(1-t)/(n+1)]/(1-tw)^2$$
$$P_T(aa) = w(1-t)[1 - tw - n\gamma w(1-t)/(n+1)]/(1'-tw)^2$$
$$+ [n\gamma w(1-t)/(n+1)]/(1-tw)$$

$$(6.3.1)$$

Hence the frequencies of AA and Aa in the next generation are given by

$$u' = p^2[1 - tw - n\gamma w(1-t)/(n+1)]/(1-tw)^2$$
$$v' = 2pq[1 - tw - n\gamma w(1-t)/(n+1)]/(1-tw)^2$$
$$- pwt[1 - tw - n\gamma w(1-t)/(n+1)]/(1-tw)^2$$
$$+ [n\gamma pw(1-t)/(n+1)]/(1-tw)$$

and the change in gene frequency by

$$\Delta p = \tfrac{1}{2}pw[t - n\gamma(1-t)/(n+1) - t^2w]/(1-tw)^2 \qquad (6.3.2)$$

This shows that at equilibrium, the phenotypic frequency may be given by

$$w_* = [t - n\gamma(1-t)/(n+1)]/t^2$$

As $n \to \infty$ this becomes the same as the equilibrium for the corresponding 'P' model with $n = 1$. The general formula, obtained by substituting the value of w_* in equation 6.3.2,

$$\Delta p = -\tfrac{1}{2}pwt^2(w - w_*)/(1-tw)^2$$

applies to both the 'PF +' model as $n \to \infty$ and the 'P' model with $n = 1$. The equilibrium for w_* is unstable because

$$d\Delta p/dp|_{p=p_*} > 0$$

When the preferred character is dominant then males with genotypes AA, Aa and aa occur with population frequencies $u(1 - s)/[1 - s(1-w)]$, $v(1-s)/[1-s(1-w)]$ and $w/[1-s(1-w)]$ giving the equations for the frequencies of matings with these males

$$P_T(AA) = u(1-s)[1-s(1-w)$$
$$+ n\alpha w/(n+1)]/[1-s(1-w)]^2$$
$$P_T(Aa) = v(1-s)[1-s(1-w)$$
$$+ n\alpha w/(n+1)]/[1-s(1-w)]^2 \qquad \Bigg\} \qquad (6.3.3)$$
$$P_T(aa) = w[1-s(1-w)$$
$$+ n\alpha w/(n+1)]/[1-s(1-w)]^2$$
$$- [n\alpha w/(n+1)]/[1-s(1-w)]$$

The change in gene frequency is then found to be

$$\Delta p = \tfrac{1}{2}pw[-s + n\alpha(1-s)/(n+1)$$
$$+ s^2(1-w)]/[(1-s(1-w)]^2 \qquad (6.3.4)$$

producing the unstable equilibrium for the dominant phenotype corresponding to that for a recessive

$$1 - w_* = [s - n\alpha(1-s)/(n+1)]/s^2$$

Generally, if the preferred character is initially at a low frequency, it will start to increase in frequency and hence ultimately reach fixation if $\gamma > t(n+1)/n(1-t)$ as a recessive and $\alpha > s(n+1)/n(1-s)$ as a dominant. As $n \to \infty$ these become the same conditions for initial increase as in the 'P' model with $n = 1$.

At any frequency, the condition for further increase of the preferred character is $\gamma > t(1-tw)(n+1)/n(1-t)$. When w is a low frequency, this gives the condition for initial increase. When w is close to unity and the character has about reached fixation, then $\gamma > t(n+1)/n$. As w increases from a low initial frequency, the value of γ that is necessary for the increase in frequency to continue is thus slightly reduced. Provided the condition for initial increase is satisfied, therefore, that increase in frequency will necessarily continue until fixation is reached. This is a consequence of the slight positive frequency-dependence in the mating success of the males who have the mating preferences in their favour.

6.4 Positive frequency-dependence and natural selection in both sexes

If the preferred character is expressed in both sexes, the females who possess it, like the males, suffer from any deleterious effects it may have on survival. They mate after natural selection with their genotypes at the same population frequencies as the males. The overall frequencies of the matings with males of each genotype are determined by the frequencies of the male genotypes and the proportions of females with preferences. These overall frequencies therefore remain the same as when selection acts only in males. They are given by equations 6.3.1 and 6.3.3; but now they are divided into the female genotypes whose frequencies have been changed from u, v and w to the same frequencies as those of the males after the natural selection has taken place.

In the model of selection for a recessive character, the frequencies of AA and Aa in the next generation are now given by

$$u' = p^2[1 - tw - n\gamma w(1-t)/(n+1)]/(1-tw)^3$$

$$v' = 2pq[1 - tw - n\gamma w(1-t)/(n+1)]/(1-tw)^3$$
$$- 2tpw[1 - tw - n\gamma w(1-t)/(n+1)]/(1-tw)^3$$
$$+ [n\gamma pw(1-t)/(n+1)]/(1-tw)^2$$

and the change in gene frequency by

$$\Delta p = pw[t - \tfrac{1}{2}n\gamma(1-t)/(n+1) - t^2 w]/(1-tw)^2 \qquad (6.4.1)$$

The unstable equilibrium frequency of the recessive phenotype is therefore

$$w_* = [t - \tfrac{1}{2}n\gamma(1-t)/(n+1)]/t^2$$

and if $\gamma > 2t(n+1)/n(1-t)$ the preferred character will always have the overall advantage and ultimately reach fixation.

Similarly, for the selection of a dominant character, we find

$$\Delta p = pw[-s + \tfrac{1}{2}n\alpha(1-s)/(n+1)$$
$$- s^2(1-w)]/[1-s(1-w)]^2 \qquad (6.4.2)$$

giving the unstable equilibrium frequency of the dominant phenotype

$$1 - w_* = [s - \tfrac{1}{2}n\alpha(1-s)/(n+1)]/s^2$$

and the condition $\alpha > 2s(n+1)/n(1-s)$ for initial increase and ultimate fixation. When selection acts in both sexes and not just in males, the mating preferences must therefore be doubled for the condition for initial increase to be satisfied.

Thus in all these models with positive frequency-dependence of the mating preferences, polymorphisms cannot be maintained by a balance between natural selection and sexual selection. The sexually advantageous character either does have an overall advantage at the start of selection or it does not. If it does, it spreads through the population to reach ultimate fixation. If it does not, it is eliminated by the adverse natural selection against it.

6.5   Negative frequency-dependent preferences for dominant and recessive characters

In this model, which I based on Spiess and Ehrman's suggestion (1978), females become less likely to mate with a particular male the more often they have encountered males with the same phenotype. Although Spiess and Ehrman's own explanation of this theory implies that females would respond in this way towards any phenotype which produced different courtship behaviour, I regard such behavioural responses as mal-adaptive and therefore implausible on general evolutionary grounds (see section 6.1). I do think, however, that such negative frequency-dependent responses might evolve for particular phenotypes or genotypes, as implied by

Averhoff and Richardson's model (1976); for several ecological factors will give rise to frequency-dependent natural selection that favours the rarer phenotypes. Batesian mimicry loses some of its advantage as particular mimics become common relative to their models (O'Donald and Pilecki, 1970); predators, who develop 'searching images' for their prey, take more of the common prey which they encounter more often (Allen and Clarke, 1968); phenotypes, which utilize different resources in their environment, will deplete less of their particular resource when rare, thus gaining an advantage over common phenotypes (Kojima, 1971). Thus, in turn, it may be advantageous for females to prefer to mate with rare phenotypes and gain an advantage by producing more and fitter offspring. But any such advantage of negative frequency-dependent mating would not evolve for rare phenotypes that were always at a selective disadvantage: as I have said, a general tendency to frequency-dependent mating would certainly produce some very mal-adaptive matings. Females would be better adapted if they evolved preferences to mate only with selectively advantageous male phenotypes. If some phenotypes had a frequency-dependent selective advantage, it would then be adaptive for females to express their preference less often as the preferred phenotype increased in frequency: negative frequency-dependence in the expression of preference would thus also be advantageous.

If females sample two available phenotypes. A and B, during encounters with $n+1$ males and express a preference either for A dependent on the frequency of B, or for B dependent on the frequency of A, then their matings with A and B males will take place at overall frequencies given by equations 3.4.1 derived in section 3.4:

$$P_T(A) = x + [\alpha y(ny+1) - \gamma x(nx+1)]/(n+1)$$

$$P_T(B) = y - [\alpha y(ny+1) - \gamma x(nx+1)]/(n+1)$$

where α and γ are the preferences for A and B. If females expressed a general tendency to mate more often with the rarer males, as suggested by Spiess and Ehrman, then the probabilities of expressing a preference would be the same for all phenotypes; for just the two phenotypes A and B, therefore, $\alpha = \gamma$; and

$$P_T(A) = x + \alpha[y + ny^2 - x - nx^2]/(n+1)$$
$$= x + \alpha(y - x)$$

$$= \alpha + x(1-2\alpha)$$

$$P_T(B) = \alpha + y(1-2\alpha)$$

These are the same frequencies of matings as those in the 'C' model of complete preferences when we put $\alpha = \gamma$. Thus, when mating preferences are roughly equal, we should expect Spiess and Ehrman's model to give a very similar fit to data on frequencies of matings as the 'C' model. I shall compare the fit of the 'C', 'P' and 'PF −' models in section 6.9 later in this chapter.

If A is dominant consisting of the genotypes AA and Aa at frequencies u and v and B is recessive consisting of the genotype aa at frequency w, then matings with these genotypes take place at frequencies

$$\left.\begin{array}{l} P_T(AA) = u + \left[\dfrac{\alpha uw(nw+1) - \gamma u(1-w)(n-nw+1)}{(n+1)(1-w)} \right] \\[2.5em] P_T(Aa) = v + \left[\dfrac{\alpha vw(nw+1) - \gamma v(1-w)(n-nw+1)}{(n+1)(1-w)} \right] \\[2.5em] P_T(aa) = w - \left[\dfrac{\alpha w(nw+1) - \gamma(1-w)(n-nw+1)}{(n+1)} \right] \end{array}\right\} \quad (6.5.1)$$

The frequencies of the genotypes AA and Aa in the next generation, and hence the change in gene frequency, are then found to be

$$u' = p^2\{1 + [\alpha w(nw+1)$$
$$- \gamma(1-w)(n-nw+1)]/(n+1)(1-w)\}$$
$$v' = 2pq\{1 + [\alpha w(nw+1)$$
$$- \gamma(1-w)(n-nw+1)]/(n+1)(1-w)\} - p[\alpha w(nw+1)$$
$$- \gamma(1-w)(n-nw+1)]/(n+1)(1-w)$$

giving the difference equation

$$\Delta p = \tfrac{1}{2}p[\alpha w(nw+1) \qquad\qquad\qquad (6.5.2)$$
$$- \gamma(1-w)(n-nw+1)]/(n+1)(1-w)$$

At equilibrium, therefore,

$$\alpha w(nw+1) = \gamma(1-w)(n-nw+1)$$

or

$$n(\alpha - \gamma)w^2 + (2n\gamma + \alpha + \gamma)w - \gamma(n+1) = 0$$

As $n \to \infty$, we obtain the equilibrium frequency

$$w_* = \sqrt{(\gamma/\alpha)}/[1 + \sqrt{(\gamma/\alpha)}] \tag{6.5.3}$$

When $n = 0$, females mate without having their preferences modified by previous encounters with courting males. The model is equivalent to the 'C' model with complete preferences, for we then have

$$P_T(A) = \alpha + x(1 - \alpha - \gamma)$$

$$P_T(B) = \gamma + y(1 - \alpha - \gamma)$$

as in the 'C' model (equations 3.1.1) and at equilibrium

$$(\alpha + \gamma)w - \gamma = 0$$

giving

$$w_* = \gamma/(\alpha + \gamma)$$

which is the equilibrium frequency given by equation 4.1.4 which .we derived in section 4.1 for the 'C' model of sexual selection for dominant and recessive characters.

At equilibrium $u_* = p_*^2$, $v_* = 2p_*q_*$ and $w_* = q_*^2$. Assuming as in previous analyses that these Hardy–Weinberg frequencies hold to good approximation in the region of the equilibrium, we can differentiate the expression for Δp, obtaining

$$d\Delta p/dp|_{p=p_*} = -p_*q_*[\alpha(2nw_* + 1)$$
$$+ \gamma(2n - 2nw_* + 1)]/(n+1)(1 - w_*)$$

As $n \to \infty$, the condition for stability becomes $\alpha w_* + \gamma(1 - w_*) > 0$ which reduces to $\sqrt{(\alpha/\gamma)} > -1$ when $w_* = \sqrt{(\gamma/\alpha)}/[1 + \sqrt{(\gamma/\alpha)}]$ and is always true when the positive square root is taken. At $n = 0$, the equilibrium is the same as in the 'C' model and always stable if $\alpha + \gamma > 0$. The general stability condition $\alpha > \gamma[1 - 2(n+1)/(2nw_* + 1)]$ must hold for all n and any equilibrium w_* when $\alpha > 0$. The polymorphic equilibria are all stable. When the mating preference is negatively frequency-dependent, then so is the mating success and a stable polymorphism therefore exists. Table 6.1 shows the selective disadvantage of phenotype B when a female preference in favour of A is either negatively frequency-dependent or complete. The selective disadvantage is always greater in the 'C' model with

Table 6.1   *Relative disadvantage of phenotype B when females prefer phenotype A*

Frequency of A	Proportion of females with preference for A					
	10% preference		30% preference		50% preference	
	'PF−'	'C'	'PF−'	'C'	'PF−'	'C'
0	1.0	1.0	1.0	1.0	1.0	1.0
0.1	0.497	0.526	0.787	0.811	0.891	0.909
0.2	0.303	0.357	0.612	0.682	0.769	0.833
0.3	0.201	0.270	0.470	0.588	0.642	0.769
0.4	0.138	0.217	0.354	0.517	0.517	0.714
0.5	0.095	0.182	0.261	0.462	0.400	0.667
0.6	0.065	0.156	0.185	0.417	0.294	0.625
0.7	0.042	0.137	0.124	0.380	0.201	0.588
0.8	0.025	0.122	0.074	0.349	0.122	0.556
0.9	0.011	0.110	0.033	0.323	0.055	0.526
1.0	0	0.1	0	0.3	0	0.5

In this table the selective disadvantage is measured by the selective coefficient calculated from the relative mating success of the phenotypes A and B. 'PF−' refers to the model with negative frequency-dependent preference and 'C' refers to the model with complete preference.

complete preference, but as expected more closely dependent on frequency in the 'PF-' model.

6.6   Negative frequency-dependent preferences for each genotype

This model of negative expression of preference for each genotype separately can be derived by assuming that females tend to become habituated against the particular genotype they prefer as its frequency increases and then mate at random with the two other genotypes. Thus among females with preferences α, β and γ for each of the genotypes AA, Aa and aa, matings take place at the following frequencies:

Females with a preference for AA males

$$P'_m = 1 - u$$

$$P_m(AA) = u(nu + 1)/(n + 1)$$

$$P_m(Aa) = nuv/(n + 1)$$

$$P_m(aa) = nuw/(n + 1)$$

Females with a preference for Aa males

$$P'_m = 1 - v$$
$$P_m(AA) = nuv/(n+1)$$
$$P_m(Aa) = v(nv+1)/(n+1)$$
$$P_m(aa) = nvw/(n+1)$$

Females with a preference for aa males

$$P'_m = 1 - w$$
$$P_m(AA) = nuw/(n+1)$$
$$P_m(Aa) = nvw/(n+1)$$
$$P_m(aa) = w(nw+1)/(n+1)$$

As in previous derivations of the 'PF$^-$' models, P'_m is the proportion of matings of females who exercise their preference for each genotype. Then we obtain the overall frequencies of the matings

$$\left.\begin{aligned}
P_T(AA) &= \alpha(n - nu + 1)/(n+1) + u[1 - (\alpha + \beta + \gamma) \\
&\quad + n(\alpha u + \beta v + \gamma w)/(n+1)] \\
P_T(Aa) &= \beta(n - nv + 1)/(n+1) + v[1 - (\alpha + \beta + \gamma) \\
&\quad + n(\alpha u + \beta v + \gamma w)/(n+1)] \\
P_T(aa) &= \gamma(n - nw + 1)/(n+1) + w[1 - (\alpha + \beta + \gamma) \\
&\quad + n(\alpha u + \beta v + \gamma w)/(n+1)]
\end{aligned}\right\} \quad (6.6.1)$$

and finally the change in gene frequency from one generation to the next according to the equation

$$\Delta p = \tfrac{1}{2}[(\alpha + \tfrac{1}{2}\beta) - n(\alpha u + \tfrac{1}{2}\beta v)/(n+1) - p(\alpha + \beta + \gamma) \\
+ np(\alpha u + \beta v + \gamma w)/(n+1)] \quad (6.6.2)$$

so that at equilibrium

$$p_* = \frac{(n+1)(\alpha + \tfrac{1}{2}\beta) - n(\alpha u_* + \tfrac{1}{2}\beta v_*)}{(n+1)(\alpha + \beta + \gamma) - n(\alpha u_* + \beta v_* + \gamma w_*)}$$

when $n = 0$, this is simply the equilibrium of the 'C' model of sexual selection for each genotype, since preferences must then always be expressed. Also at equilibrium $u_* = p_*^2$, $v_* = 2p_*q_*$ and $w_* = q_*^2$ and the general equation for the equilibrium frequency of p_* can therefore be expressed as a cubic equation in p_*.

Two special cases are worth detailed analysis. If preferences only favour heterozygotes so that $\alpha = \gamma = 0$, then as $n \to \infty$ the cubic equation can be reduced to

$$4p_*^3 - 6p_*^2 + 4p_* - 1 = 0$$

or $$(2p_* - 1)(2p_*^2 - 2p_* + 1) = 0$$

showing that $p_* = \frac{1}{2}$ represents the polymorphic equilibrium. Since

$$\Delta p = \tfrac{1}{4}\beta(1 - v)(1 - 2p)$$

therefore

$$d\Delta p / dp|_{p = \frac{1}{2}} = -\tfrac{1}{4}\beta$$

and as expected the equilibrium is always stable.

If preferences favour the two homozygotes so that $\alpha = \gamma$, $\beta = 0$, then as $n \to \infty$

$$(2p_* - 1)(p_*^2 - p_* - 1) = 0$$

$$p_* = \tfrac{1}{2}$$

$$\Delta p = \tfrac{1}{2}\alpha[1 - u - p(1 + v)]$$

$$= \tfrac{1}{2}\alpha(1 - p - 3p^2 + 2p^3)$$

$$d\Delta p / dp|_{p = \frac{1}{2}} = -(5/4)\alpha$$

and this equilibrium is also stable in spite of the heterozygous disadvantage. The stability is of course the result of the strong negative frequency-dependence in the model. More generally, if $\beta = 0$, we have at equilibrium $(n \to \infty)$

$$(\alpha + \gamma)p_*^3 - (\alpha + 2\gamma)p_*^2 - \alpha p_* + \alpha = 0 \tag{6.6.3}$$

giving stable equilibria in spite of the complete absence of heterozygote preference. In discussions of the evolution of mating preferences for separate genotypes (Chapter 8, sections 8.3 and 8.4) separate preferences for heterozygotes are often eliminated, leaving only the two homozygotes as the preferred genotypes. I have made this assumption in using equation 6.6.3 to predict equilibrium frequencies from estimates of mating preferences obtained from experiments to test the relative mating success of different homozygotes at varying frequencies (see section 6.9).

6.7 Negative frequency-dependence and natural selection in males

If phenotype A is dominant (genotypically AA or Aa) and B is recessive (genotypically aa) then the females mate with each genotype at overall frequencies given by equations 6.5.1 in section 6.5. After natural selection of the males has taken place, however, the frequencies of the genotypes AA, Aa and aa are no longer u, v and w; but as in other models with natural selection they become either $u/(1-tw)$, $v/(1-tw)$ and $w(1-t)/(1-tw)$ if B is the preferred phenotype with selective disadvantage t, or $u(1-s)/[1-s(1-w)]$, $v(1-s)/[1-s(1-w)]$ and $w/[1-s(1-w)]$ if A is the preferred phenotype with selective disadvantage s.

When females prefer the recessive phenotype B, then by substituting the frequencies after natural selection for u, v and w, we obtain the equations for the frequencies of matings

$$P_T(AA)=u[1-tw-\gamma(1-tw+n-nw)/(n+1)]/(1-tw)^2$$
$$P_T(Aa)=v[1-tw-\gamma(1-tw+n-nw)/(n+1)]/(1-tw)^2$$
$$P_T(aa)=w(1-t)[1-tw-\gamma(1-tw+n-nw)/(n+1)]/(1-tw)^2$$
$$+[\gamma(1-tw+n-nw)/(n+1)]/(1-tw) \qquad (6.7.1)$$

After considerable algebraic reduction, the change in gene frequency can be shown to be

$$\Delta p=\tfrac{1}{2}p[-\gamma+w(n\gamma+nt+t+\gamma t)/(n+1)-t^2w^2]/(1-tw)^2 \qquad (6.7.2)$$

This gives a quadratic equation to be solved for the equilibrium frequency, w_*, of the preferred phenotype. As $n\to\infty$, we get the equation

$$\Delta p=\tfrac{1}{2}p[-\gamma+w(\gamma+t)-t^2w^2]/(1-tw)^2$$

and when $n=0$

$$\Delta p=\tfrac{1}{2}p[-\gamma+tw]/(1-tw)$$

which is the same as equation 4.3.3 for the case when preferences are complete, as we should expect.

Approximately, when t is small we may put

$$w_*=\gamma(n+1)/(n\gamma+nt+t+\gamma t) \qquad (6.7.3)$$

Differentiating the expression for Δp near the equilibrium and using the approximate value of w_* given by 6.7.3 we find that

$$d\Delta p/dp|_{p=p_*} = -p_* q_* (n\gamma + nt + t + \gamma t)/(n+1)(1-tw_*)^2$$

hence showing that the polymorphic equilibrium is always stable for biologically meaningful values of γ and t. Obviously, when w is small, Δp will always be negative and the character must always increase in frequency to the stable equilibrium point.

Equation 6.7.2 for Δp can be written in the form

$$\Delta p = \tfrac{1}{2} p(w - w_*)[n\gamma + nt + t + \gamma t - t^2(n+1)(w+w_*)]/[(n+1)(1-tw)^2]$$

If t is small and t^2 negligible (a necessary condition for assuming that Δp is small and can be treated as a differential), then approximately

$$\Delta p = \tfrac{1}{2}(1-q)(q-q_*)(q+q_*)(n\gamma + nt + t + \gamma t)/(n+1)$$

Since $\Delta p = -\Delta q$ we obtain on integration of the corresponding differential equation

$$\left(\frac{1-q_0}{1-q_T}\right)^{2q_*} \cdot \left(\frac{q_T - q_*}{q_0 - q_*}\right)^{(1+q_*)} \cdot \left(\frac{q_0 + q_*}{q_T + q_*}\right)^{(1-q_*)}$$

$$= \exp[-Tq_*(1-q_*^2)(n\gamma + nt + t + \gamma t)/(n+1)] \qquad (6.7.4)$$

since it can be shown that

$$\int \frac{dq}{(1-q)(q-q_*)(q+q_*)}$$

$$= \frac{1}{2q_*(1-q_*^2)} \int \left(\frac{2q_*}{1-q} + \frac{1+q_*}{q-q_*} - \frac{1-q_*}{q+q_*}\right) dq$$

and the integration then follows.

When females prefer the dominant phenotype A which has a selective disadvantage s, then the frequencies of the matings

$$P_T(AA) = u(1-s)\left[1 - s + sw \right.$$

$$\left. + \frac{\alpha w(1-s+sw+nw)}{(n+1)(1-s)(1-w)}\right] / (1-s+sw)^2$$

$$P_T(Aa) = v(1-s)\left[1-s+sw\right.$$

$$\left. + \frac{\alpha w(1-s+sw+nw)}{(n+1)(1-s)(1-w)}\right] / (1-s+sw)^2 \qquad (6.7.5)$$

$$P_T(aa) = w\left[1-s+sw\right.$$

$$\left. + \frac{\alpha w(1-s+sw+nw)}{(n+1)(1-s)(1-w)}\right] / (1-s+sw)^2$$

$$\left. - \left[\frac{\alpha w(1-s+sw+nw)}{(n+1)(1-s)(1-w)}\right] / (1-s+sw)\right.$$

can be shown, after heavy algebra, to give a change in gene frequency,

$$\Delta p = \tfrac{1}{2}pw[\alpha - (1-w)(n\alpha + ns + s + \alpha s)/(n+1)$$

$$+ s^2(1-w)^2] / \{(1-w)[1-s(1-w)]^2\} \qquad (6.7.6)$$

Thus the equilibrium frequency, $1-w_*$, of the preferred phenotype, corresponds to that of the recessive. Approximately, if s is small,

$$1-w_* = \alpha(n+1)/(n\alpha + ns + s + \alpha s) \qquad (6.7.7)$$

and equation 6.7.6 then becomes

$$\Delta p = \tfrac{1}{2}pw(w-w_*)(n\alpha + ns + s + \alpha s)/[(n+1)(1-w)]$$

$$= \tfrac{1}{2}q^2(q^2 - q_*^2)(n\alpha + ns + s + \alpha s)/[(n+1)(1+q)]$$

Treating this equation as a differential and integrating we obtain

$$-\tfrac{1}{2}T(n\alpha + ns + s + \alpha s) = \int_{q_0}^{q}\left[\frac{(1+q)\,dq}{q^2(q^2 - q_*^2)}\right]$$

$$= \frac{1}{2q_*^2}\int_{q_0}^{q_T}\left(\frac{1+q_*}{q_*(q-q_*)} - \frac{1-q_*}{q_*(q+q_*)} - \frac{2}{q^2} - \frac{2}{q}\right)dq$$

and therefore

$$(1+q_*)\log_e\left(\frac{q_T - q_*}{q_0 - q_*}\right) + (1-q_*)\log_e\left(\frac{q_0 + q_*}{q_T + q_*}\right) + 2q_*\log_e\left(\frac{q_0}{q_T}\right)$$

$$+ \frac{2q_*}{q_T} - \frac{2q_*}{q_0} = -q_*^3 T(n\alpha + ns + s + \alpha s)/(n+1) \qquad (6.7.8)$$

This expression is probably too cumbersome to be of use in fitting data on changes in gene frequency. It might be simpler to use the exact recurrence equations

$$u' = p^2(1-s)\left[1-s+sw+\frac{\alpha w(1-s+sw+nw)}{(n+1)(1-s)(1-w)}\right]\bigg/(1-s+sw)^2$$

$$v' = \left[\frac{2pq(1-s)+spw}{(1-s+sw)^2}\right]\cdot\left[1-s+sw+\frac{\alpha w(1-s+sw+nw)}{(n+1)(1-s)(1-w)}\right]$$

$$-\frac{\alpha pw(1-s+sw+nw)}{(n+1)(1-s)(1-w)(1-s+sw)}$$

to obtain numerical values for the changes in gene frequency, fitting data by the trial-and-error method outlined in Chapter 4, section 4.7.

6.8   Negative frequency-dependence and natural selection in both sexes

As in the derivation of previous models with natural selection acting against the preferred character in both males and females, the overall frequencies of matings given by equations 6.7.1 and 6.7.5 are now divided among females who have the same genotypic frequencies after natural selection as the males. When females prefer the recessive phenotype, B, we can derive recurrence equations for the AA and Aa genotypes in the form

$$u' = p^2[1-tw-\gamma(1-tw+n-nw)/(n+1)]/(1-tw)^3$$

$$v' = (2pq-2tpw)[1-tw-\gamma(1-tw+n-nw)/(n+1)]/(1-tw)^3$$

$$+\gamma p[(1-tw+n-nw)/(n+1)]/(1-tw)^2$$

and it then follows that

$$\Delta p = p[-\tfrac{1}{2}\gamma+w(\tfrac{1}{2}n\gamma+nt+t+\tfrac{1}{2}\gamma t)/(n+1) \tag{6.8.1}$$
$$-t^2w^2]/(1-tw)^2$$

For small t, therefore, the equilibrium is given approximately by

$$w_* = \tfrac{1}{2}\gamma(n+1)/(\tfrac{1}{2}n\gamma+nt+t+\tfrac{1}{2}\gamma t) \tag{6.8.2}$$

and approximately

$$\Delta p = p(q^2-q_*^2)(\tfrac{1}{2}n\gamma+nt+t+\tfrac{1}{2}\gamma t)/(n+1)$$

The integration is similar to that carried out for equation 6.7.4. We find

$$\left(\frac{1-q_0}{1-q_T}\right)^{2q_*}\cdot\left(\frac{q_T-q_*}{q_0-q_*}\right)^{(1+q_*)}\cdot\left(\frac{q_0+q_*}{q_T+q_*}\right)^{(1-q_*)}$$
$$=\exp[-2Tq_*(1-q_*^2)(\tfrac{1}{2}n\gamma+nt+t+\tfrac{1}{2}\gamma t)/(n+1)] \qquad (6.8.3)$$

As in all previous models, with natural selection in both sexes, the preference must be doubled to produce the same equilibrium frequency and the same condition for initial increase as when natural selection acts only in the males.

When females prefer the dominant phenotype A, we obtain the recurrence equations

$$u'=p^2(1-s)^2\left[1-s+sw+\frac{\alpha w(1-s+sw+nw)}{(n+1)(1-s)(1-w)}\right]\bigg/(1-s+sw)^3$$

$$v'=\left[\frac{2pq(1-s)^2+2spw(1-s)}{(1-s+sw)^3}\right]\left[1-s+sw\right.$$
$$\left.+\frac{\alpha w(1-s+sw+nw)}{(n+1)(1-s)(1-w)}\right]-\frac{\alpha pw(1-s+sw+nw)}{(n+1)(1-w)(1-s+sw)^2}$$

and therefore

$$\Delta p=pw[\tfrac{1}{2}\alpha-(1-w)(\tfrac{1}{2}n\alpha+ns+s+\tfrac{1}{2}\alpha s)/(n+1)$$
$$+s^2(1-w)^2]/[(1-w)(1-s+sw)^2] \qquad (6.8.4)$$

with an approximate equilibrium frequency of A at

$$1-w_*=\tfrac{1}{2}\alpha(n+1)/(\tfrac{1}{2}n\alpha+ns+s+\tfrac{1}{2}\alpha s) \qquad (6.8.5)$$

and a cumbersome equation, similar to 6.7.8, for the gene frequency after T generations of selection:

$$(1+q_*)\log_e\left(\frac{q_T-q_*}{q_0-q_*}\right)+(1-q_*)\log_e\left(\frac{q_0+q_*}{q_T+q_*}\right)+2q_*\log_e\left(\frac{q_0}{q_T}\right)$$
$$+\frac{2q_*}{q_T}-\frac{2q_*}{q_0}=-2q_*^3\ T(\tfrac{1}{2}n\alpha+ns+s+\tfrac{1}{2}\alpha s)/(n+1) \qquad (6.8.6)$$

6.9   Fitting the models to data

The 'PF$-$' models with negative frequency-dependent preferences can be fitted to data on numbers of matings with different male phenotypes in the same way as the 'C' and 'P' models were fitted. Using the same data from experiments carried out by Ehrman and Spiess, the fit of all three models 'C', 'P' and 'PF −' can be compared. This comparison is particularly interesting because certain of the models will be excluded if other models provide a significantly better fit. It may therefore be possible to make inferences about the mechanism of sexual selection *Drosophila*. The fitting of model 'PF −' represents the negative frequency-dependence in female preference implied by Spiess and Ehrman's own suggestion that females become habituated to males they encounter and mate with different males who break the habituation. If this were a general tendency to habituation against any phenotype, as Spiess and Ehrman imply, then the same number of encounters with any phenotype would produce the same proportion of females who had become habituated, or had the same chance of becoming habituated, implying $\alpha = \gamma$ when only two phenotypes of males were being chosen as mates. The parameter α would either be the proportion of the females habituated, or the chance of their becoming habituated when one of the phenotypes had been encountered: none of these habituated females would then mate with the males possessing the phenotype they had become habituated to.

In the general 'PF −' model in which different proportions of females may become habituated to the different males, these proportions are equivalent to the mating preferences to be estimated from the data. In the experiments on mating choice between males, females chose their mates from fixed proportions of males, not from binomial samples. Thus n is not a parameter to be estimated or inserted in the model: the number of each of the two genotypes of males was fixed for each frequency at which female mating choice was tested. The appropriate model is therefore the 'PF −' model in the case when $n \to \infty$. The expected frequencies of matings are then given by

$$P_T(A) = x + \alpha y^2 - \gamma x^2$$
$$P_T(B) = y - \alpha y^2 + \gamma x^2$$

In fitting the 'P' models, however, as shown in Chapter 5, section 5.5, n is the number of encounters required to provide the extra stimulation for a female to mate with a male who is not her preferred choice. A female is assumed to sample the males successively until she is sufficiently stimulated to mate: each male who has been encountered remains in the pool of available males so that the probability of sampling a particular male phenotype stays the same and is fixed in each experiment at the particular frequency being tested. The parameter n can vary although the frequency of males cannot. Values of the log likelihood were calculated for a wide range of values of n, and at each value the maximum likelihood estimates of α and γ were found. That value of n that gave the highest overall likelihood has been taken from tables 5.8 and 5.9 given in section 5.5. The estimates of α and γ corresponding to the maximum likelihood estimate of n are shown in table 6.2 together with the estimates obtained by fitting the 'C' and 'PF −' models to the same data. Also given in the table are the equilibrium frequencies calculated from the estimates of the preferences on the assumption that α and γ are separate preferences for the particular homozygous genotypes tested in the experiments. As expected, if the estimates of the preferences for the two genotypes are roughly equal, the 'C' and 'PF −' models give a very similar fit: at equal frequencies of the preferences and with fixed proportions of males for the expression of female choice, the expectations of the two models are identical. The data cannot then refute either one or other of the models.

In one set of data (Ehrman's positively selected line of *D. pseudoobscura* with karyotypes *CH* and *AR*), the 'PF −' model gives a significantly better fit than the other two models. However, in Ehrman's experiments with *CH* and *AR*, significant heterogeneity is left after fitting each of the models. The heterogeneity affects certain frequencies in such a way that no simple model of frequency-dependent mating could account for it: in these data, at some of the intermediate frequencies, the rare males gain a much greater relative mating success than at the extreme frequencies when the same males are much rarer and should therefore be at relatively greater advantage. I suggested that this might be explained by heterogeneity in female preferences between the experiments (O'Donald, 1977a): if females vary in their responses, their average response may also vary by sampling between the females chosen to test for mating preferences at different male frequencies. The variance in the observed numbers of matings will then be greater than the binomial variance. Values of χ^2, calculated on the assumption of binomial variation, will therefore be inflated.

Generally, however, the 'P' model gives the best fit. The fit of the 'P' model to the data of Spiess (1968) and Spiess and Spiess (1969)

Table 6.2 *Comparison of models of complete, partial and frequency-dependent preferences fitted to data of Spiess and Ehrman*

Data fitted	Complete preference model 'C'	Partial preference model 'P'	Frequency-dependence model 'PF–'
Spiess (1968) on D. pseudoobscura karyotypes AR (pref. α) and PP (pref. γ)	$\chi_9^2 = 67.105$ $\hat\alpha = 0.1380$ $\hat\gamma = 0.0210$	$\chi_9^2 = 67.105$ $\hat\alpha = 0.4184$ $\hat\gamma = 0.1919$ $\hat n = 3$	$\chi_9^2 = 67.105$ $\hat\alpha = 0.1540$ $\hat\gamma = 0.0062$
	$\log L = -795.151$ $\chi_7^2 = 8.672$ $p_* = 0.868$	$\log L = -793.584$ $\chi_6^2 = 5.620$ p_* unstable	$\log L = -797.424$ $\chi_7^2 = 13.117$ $p_* = 0.865$
Spiess and Spiess (1969) on D.persimilis strains Hu (pref. α) and WW (pref. γ)	$\chi_9^2 = 101.707$ $\hat\alpha = 0.1656$ $\hat\gamma = 0.1557$	$\chi_9^2 = 101.707$ $\hat\alpha = 0.4484$ $\hat\gamma = 0.4402$ $\hat n = 5$	$\chi_9^2 = 101.707$ $\hat\alpha = 0.1682$ $\hat\gamma = 0.1530$
	$\log L = -457.467$ $\chi_7^2 = 13.692$ $p_* = 0.515$	$\log L = -453.946$ $\chi_6^2 = 6.140$ p_* unstable	$\log L = -457.452$ $\chi_7^2 = 13.662$ $p_* = 0.514$
Ehrman (1967) on D.pseudoobscura karyotypes CH (pref. α) and AR (pref. γ) (negatively selected line)	$\chi_9^2 = 170.080$ $\hat\alpha = 0.2426$ $\hat\gamma = 0.2917$	$\chi_9^2 = 170.080$ $\hat\alpha = 0.4499$ $\hat\gamma = 0.5232$ $\hat n = 6$	$\chi_9^2 = 170.080$ $\hat\alpha = 0.2402$ $\hat\gamma = 0.2932$
	$\log L = -672.860$ $\chi_7^2 = 22.044$ $p_* = 0.454$	$\log L = -670.578$ $\chi_6^2 = 17.537$ p_* unstable	$\log L = -673.532$ $\chi_7^2 = 23.391$ $p_* = 0.470$
Ehrman (1967) on D.pseudoobscura karyotypes CH (pref. α) and AR (pref. γ) (positively selected line)	$\chi_9^2 = 244.051$ $\hat\alpha = 0.2167$ $\hat\gamma = 0.3277$	$\chi_9^2 = 244.051$ $\hat\alpha = 0.2311$ $\hat\gamma = 0.3411$ $\hat n = 25$	$\chi_9^2 = 244.051$ $\hat\alpha = 0.1692$ $\hat\gamma = 0.3731$
	$\log L = -657.095$ $\chi_7^2 = 30.460$ $p_* = 0.398$	$\log L = -656.996$ $\chi_6^2 = 30.323$ $p_* = 0.401$	$\log L = -654.448$ $\chi_7^2 = 25,645$ $p_* = 0.384$
Ehrman (1970) on D.pseudoobscura homozygotes or (*pref.* α) and pr (pref. γ)	$\chi_{20}^2 = 64.132$ $\hat\alpha = 0.2436$ $\hat\gamma = 0.2588$	$\chi_{20}^2 = 64.132$ $\hat\alpha = 0.2673$ $\hat\gamma = 0.2819$ $\hat n = 10$	$\chi_{20}^2 = 64.132$ $\hat\alpha = 0.2363$ $\hat\gamma = 0.2674$
	$\log L = -852.879$ $\chi_{18}^2 = 30.823$ $p_* = 0.485$	$\log L = -852.855$ $\chi_{17}^2 = 30.743$ $p_* = 0.478$	$\log L = -852.845$ $\chi_{18}^2 = 30.763$ $p_* = 0.481$

Values p_* have been calculated on the assumption that α and γ are separate preferences for the two homozygous genotypes. For the 'C' model, p_* has been calculated by the expression $\alpha/(\alpha+\gamma)$. For the 'PF–' model, p_* has been obtained as the solution of the cubic equation 6.6.3:

$$(\alpha+\gamma)p_*^3 - (\alpha+2\gamma)p_*^2 - \alpha p_* + \alpha = 0$$

Values of p_* for the 'P' model were obtained by iteration of the recurrence equations of the genotypic frequencies.

is very significantly better than the fit of the 'C' and 'PF —' models to these data. Relatively small values of $\hat{n}$ give the mximum likelihood, but the proportions of females with preferences is greatly increased over those with preferences in the other models. This is because a much smaller proportion of females express their preferences in the 'P' models when only a few encounters with the 'wrong' males are sufficient to stimulate a female to mate with them. At the smaller values of n, the frequency-dependence of the mating success is reduced in comparison with the 'C' and 'PF —' models, and in the data of Spiess and Spiess, the reduced level of frequency-dependence determines the better fit to the 'P' models at small n values.

The equilibrium frequencies given in table 6.2 are calculated on the assumption that α and γ are separate preferences for the two homozygous genotypes. I have assumed in these calculations that there are no preferences for the heterozygotes. Computer simulations of the evolution of separate preferences show that any preference for the heterozygotes is virtually eliminated for most parameter values while preferences for the homozygotes increase in frequency. (The computer models and the inferences that can be drawn from them are discussed in detail in Chapter 8.) In the absence of data on comparative mating success of heterozygotes and homozygotes, some assumption must be made about the preference for heterozygotes before the equilibrium gene frequency can be calculated. On the basis of the computer simulations, I took the preference to be zero, rather than give it some arbitrary value. Population cage experiments could be used to determine actual equilibrium frequencies and thus provide a simple independent test of the models. The predicted equilibrium frequency is calculated only from the estimates of the preferences: it makes no allowance for natural selection, which would therefore upset the prediction. If fitness parameters could be estimated, they could be incorporated in the model and used to make a better prediction of the equilibrium. I have already given the analytical results for the 'C', 'P' and 'PF' models when a single preferred character has a deleterious effect on survival. Numerical solutions of the recurrence equations of genotypic frequencies could easily be obtained for more general models with more than one preferred character. Alternatively, given a preference estimated for a particular character, the relative fitness could then be found from both the equilibrium frequency and the rate of approach to it.

In the 'P' models, when no females prefer heterozygotes, equilibria are only stable at higher values of n. The 'P' models have, by then, closely converged on the 'C' models. Since in general, population cage experiments show that stable polymorphisms of the *Drosophila* karyotypes can be maintained, the unstable equilibria in the 'P' models seem to be a decisive refutation. But these equilibria have been calculated on the assumption that females prefer homozygotes but not heterozygotes. This always gives rise to unstable equilibria at the lower values of n: the equilibria are unstable if sexual selection favours the two homozygotes or if sexual selection favours one homozygote and natural selection favours the other. If some females did prefer heterozygotes, the instability of the equilibria might then give way to stability.

7

Sexual selection in monogamous species

7.1 Darwin's theory of sexual selection in monogamous birds

When monogamy is the rule, the sexually more favoured individuals gain an advantage from their better prospects of mating only if they produce more offspring than the less favoured individuals. In any model of mating behaviour, the sexually more favoured individuals will find mates sooner than the less favoured. The possibility of sexual selection in a monogamous species then depends on a relationship between breeding time and reproductive success as originally suggested by Darwin (1871). Darwin thought that earlier breeding would be correlated with increased reproductive success since those females in better nutritional condition after the winter would develop sooner and produce larger numbers of better nourished offspring. Males mating with the earlier females would therefore contribute more offspring to the next generation. Darwin stated his theory in these words:

Let us take any species, a bird for instance, and divide the females inhabiting a district into two equal bodies, the one consisting of the more vigorous and better nourished individuals, and the other of the less vigorous and healthy. The former, there can be little doubt, would be ready to breed in the spring before the others ... There can also be no doubt that the most vigorous, best nourished and earliest breeders would on an average succeed in rearing the largest number of fine offspring. The males, as we have seen, are generally ready to breed before the females; the strongest, and with some species the best armed of the males, drive away the weaker; and the former would then unite with the more vigorous and better nourished females, because they are the first to breed. Such vigorous pairs would surely rear a larger number of offspring than the retarded females, which would be compelled to unite with the conquered and less powerful males, supposing the sexes to be numerically equal; and this is all that is wanted to add, in the course of successive generations, to the size, strength and courage of the males, or to improve their weapons.

Fisher (1930) pointed out that such a correlation would have to be entirely environmental, and not at all genetic, if selection were not to produce an increasingly early breeding season. He gave a numerical example of how such a correlation could arise. Darwin was certainly right in his assumption that reproductive success declines as the breeding season advances. Many non-passerine birds show this effect (Lack, 1968). O'Donald (1972a, b, 1973a, 1974, 1976) used the actual relationship of breeding time and reproductive success in a seabird, the Arctic Skua, to compute the selective advantage gained as a result of a mating advantage of particular male phenotypes. The rate of sexual selection can then be found for an allele that determines mating success. In fact, the males' selective advantage does not necessarily depend on reproductive success declining from the beginning to the end of the breeding season. If reproductive success is greatest in the middle of the breeding season and decreases among pairs that breed either earlier or later, selection for the preferred males can still take place: a selective advantage is gained either if the frequency of preferred males is high, or if the frequency of females with preferences is low (O'Donald, 1973a). If the frequency of the preferred males is greater than 50 per cent, the males who are not the females' preference will necessarily be left unmated until after the mean breeding date, and they will therefore have a lower average reproductive success than the others. If only a small proportion of females have preferences, then the preferred males generally find mates around the mean breeding date, while a higher proportion of the remaining males are left to mate towards the end of the breeding season with a lower average reproductive success than the preferred males.

As mentioned briefly in Chapter 2, section 2.3, the Arctic Skua is polymorphic in its plumage: it has a pale non-melanic phenotype, and intermediate and dark, melanic phenotypes. These phenotypes are genetically determined by two alleles: the intermediates are mostly the heterozygotes, but they overlap phenotypically with the dark homozygotes. Generally the melanics have a greater reproductive success than the non-melanics. In accordance with Darwin's theory, this difference in reproductive success depends partly on the melanic males' earlier pairing during the breeding season. A rather similar polymorphism has been observed in the feral pigeon which has a wild-type 'blue' phenotype and melanic 'blue checker' and 'dark blue-checker' phenotypes in urban populations (Murton, Westwood and Thearle, 1973; discussed in detail in section 2.3). In both the Arctic Skua and feral pigeon, a proportion of the matings are assortative. The assortative mating preferences can be estimated directly from the numbers of matings between the different phenotypes (Davis and O'Donald, 1976a, b). In the Arctic Skua, the data on the distributions of breeding times of the melanic and non-melanic males can be used to give independent estimates of the mating preferences. The more females that prefer melanic males, the earlier the melanic males will mate compared to the non-melanics: towards the end of the breeding season only the non-melanics are left for the females to mate with. Thus the proportions of females with preferences determine the distributions of breeding times of the melanic and non-melanic males. Values of the mating preferences can thus be found that give the closest fit to the actual distributions of the males' breeding times. In the Arctic Skua, intermediate melanics mate assortatively as well as breeding earlier in the season than non-melanics. The preference for intermediates estimated from the distributions of the breeding times agrees closely with the estimate obtained from the proportion of assortative matings of intermediates.

The differences between melanic and non-melanic males in their breeding times are only observed when the males are mating for the first time or when they have changed their mate: males who have bred with the same female in previous years show no difference in breeding times. Thus melanic Arctic Skuas show no general earlier onset of reproductive activity. Melanic pigeons, however, show an altered photoperiodism with earlier recrudescence of the gonads and a lengthened breeding season compared to the non-melanics. In the Arctic Skuas on the contrary, melanic males who have mated with a new female actually have a shorter breeding season than non-melanics. This is a consequence of the fact that on average the melanic males find their mates sooner than the non-melanic males: a considerable proportion of the non-melanics are left to breed towards the end of the season, which is thus extended by a long tail of late breeding non-melanics. The models of female preference predict exactly this effect. Data are not available on the breeding dates of the melanic and non-melanic pigeons. If the melanic pigeons have mating preferences in their favour as well as their altered photoperiodism, we should then expect that melanic males, who were paired to new females, would find mates sooner on average than the same melanics paired with females they had bred with in previous years; and some non-melanics should be left unmated until much later in the breeding season than any of the melanics. Data of breeding times of new and old pairs of melanic and non-melanic pigeons would therefore provide an independent test of the hypothesis that the assortative mating in pigeons is a consequence of female preference.

7.2 Models of mating preferences with monogamy

·O'Donald (1976) described a number of specific models in terms of mating preferences for the different male phenotypes of the Arctic Skua. In these models selection of one or more of the phenotypes can either take place indiscriminately, or the selection of the phenotypes can take place separately, or there can be combinations of these possibilities.

In monogamous mating systems allowance must also be made in the models for the order in which the matings take place: a male who has once mated is removed from the pool of males which the females can choose from. The sequence of preferential and random matings thus becomes important. Different proportions of the males are available for the females to choose from if the preferential matings follow rather than precede, the random matings. This alters the estimates of the mating preferences required to produce the actual distributions of breeding times.

Four basic models describe how the females' mating preferences determine the probabilities with which the dark, intermediate and pale males of the Arctic Skuas obtain their mates during the breeding season. In the simplest model, which I called *Model* 1 (O'Donald, 1976), some of the females prefer to mate either with dark or with intermediate males. The remaining females mate at random with dark, intermediate and pale males. The females with the preferences are also prepared to mate with pale males if no darks and intermediates remain unmated. I used this model (O'Donald, 1973a) to analyse the theoretical effects of sexual selection. The selective coefficients were determined by the relationship between breeding time and breeding success. They are frequency-dependent and may either increase or decrease with frequency: they increase with frequency when most of the females have the preference; when less than about 30 per cent of the females have the preference, they decrease. Selection for the melanic dark and intermediate males corresponds to the model with polygyny in which a dominant gene determines the preferred phenotype.

In *Model* 2, the females with the preference mate with dark males if they can: when all the dark males have found mates, they mate with the intermediate males; they only mate with pale males if no darks and intermediates are available. The females with no preference mate at random with available males. In *Model* 3, there are two groups of females with different preferences: some of them, a proportion α, prefer to mate only with dark males; others, a proportion, β, prefer either dark or intermediate males indiscriminately; the remaining females, the proportion $1 - \alpha - \beta$, mate at random. The females with preferences also mate at random if no males with the preferred phenotype remain without mates. Thus when $\alpha = 0$, model 3 is identical to model 1: the dark allele is then dominant in its effect on mating preference. Finally, in *Model* 4, α of the females prefer dark males and β prefer intermediate males. This is a particular case of the more general model in which there is a separate preference for each phenotype (O'Donald, 1974).

In fitting these models of the mating preferences to the data of the Arctic Skua on Fair Isle, O'Donald, Wedd and Davis (1974) considered only those models in which preferential matings take place before random matings (called Models 1P, 2P, 3P and 4P). When the random matings precede the preferential matings (O'Donald, 1976) we obtain Models 1R, 2R, 3R and 4R. Preferential and random matings may also occur simultaneously giving rise to Models 1S, 2S, 3S and 4S. It has not been possible to derive the distributions of breeding times from these models mathematically. But, given particular values of the preferences, the distributions can be found by computer simulation.

In all the models, the females become willing to mate during successive intervals in the breeding season. The proportion of pairs actually breeding in a particular interval is then assumed to be the proportion of females who came into breeding condition and mated in that interval.

Table 7.1 gives the actual distribution of breeding dates of melanic and non-melanic male Arctic Skuas who are mating with a particular female for the first time (O'Donald and Davis, 1977). As stated in the previous section, melanic males generally have the earlier breeding dates, but the differences in breeding dates between melanics and non-melanics disappears among males who are breeding with the same female they bred with in previous years: the difference is found only in new pairs showing that it is determined by the process of mate selection (see section 7.1).

Table 7.1   *Distributions of breeding dates of male melanic and non-melanic Arctic Skuas in new pairs*

Breeding dates in weekly intervals	Numbers of males breeding in each interval		
	Dark	Intermediate	Pale
10–16 June	2	2	1
17–23 June	9	33	4
24–30 June	21	46	9
1–7 July	15	33	13
8–14 July	5	21	12
15–21 July	3	7	5
Totals	55	142	44

7.3   Fitting the Arctic Skua data to the models

O'Donald and Davis (1977) fitted the data on the distributions of breeding dates to the different models. For given proportions of females with preferences, computer simulation of the models gives the probabilities that dark, intermediate and pale males will breed in each successive interval of the breeding season. These intervals may be the weeks shown in table 7.1 or each successive day. If probabilities for each day were calculated, they were lumped to give probabilities corresponding to each of the weeks in the table. The log likelihoods were then calculated for arbitrary values of the mating preferences. Random perturbations were introduced into the values of the mating preferences and the log likelihoods calculated again, the process continuing until a higher log likelihood had been found. As in the fitting of the models with polygynous matings, this was repeated with smaller and smaller random perturbations until the maximum log likelihood of the model had been reached. The maximum likelihood estimates of the parameters of the models are given in table 7.2. The results shown for Models 4P, 4R and 4S are only given for the case when $\gamma = 0$. When this is not the case, the likelihoods are very slightly increased, but the general effect is negligible. Model 4P has the highest likelihood. In general the P models have the highest likelihoods, followed by the S models. However the differences in the log likelihoods of all the models 3 and 4 are small. Only Models 2P and 2R have log likelihoods more than two units of support below the maximum and these models can therefore be rejected.

Table 7.2   *Maximum likelihood estimates of the parameters of the models for the data of table 7.1*

Model	Parameters of female preference		Log likelihood (to base *e*)
	α	β	
1P	–	0.245	– 336.509
1R	–	0.246	– 337.695
1S	–	0.337	– 337.004
2P	0.065	–	– 340.441
2R	0.043	–	– 340.191
3P	0.024	0.402	– 336.429
3R	0.023	0.221	– 337.535
3S	0.023	0.313	– 336.863
4P	0.137	0.292	– 336.302
4R	0.084	0.165	– 337.420
4S	0.111	0.228	– 336.765

In models, 2, 3 and 4, α is the proportion of females preferring dark melanic males. β is the proportion of females preferring dark and intermediate melanics indiscriminately in models 1 and 3, or the proportion of females preferring only intermediate melanics in model 4. In model 2, α prefer first dark melanics, but then intermediate melanics if no darks are left for mating.

Table 7.1 now includes all data that have been collected up to 1976 from a colony of Arctic Skuas on the island of Fair Isle in Shetland. Data previously analysed (O'Donald, Wedd and Davis, 1974; O'Donald, 1976) had been collected up to 1973 and gave the R models the highest likelihoods. But the differences in the likelihoods between P, R and S models were very small. The data, which are now complete up to 1976, show no heterogeneity in the estimates of the mating preferences between the two periods in which the Arctic Skuas on Fair Isle have been studied. The intervals into which the breeding season is divided are also part of the model. Mate selection may take place within groups of males and females as suggested in Chapter 3. If some females had a lower threshold of response to melanic male phenotypes but not to the others, they would choose to mate with any melanic male who remained unmated in the interval in which the females became ready to breed. The remaining females, and those with no males they preferred to mate with, would then mate at random with the remaining males. This is exactly the sequence of events in the P models, which provide the best fit to the data on distributions of breeding times in the Arctic Skua. The R models may apply to sexual selection caused by variation in intensity of male courtship: females with low thresholds would respond quickly to males with both high and low levels of courtship behaviour and mate more or less at random between the males; females with higher thresholds would respond relatively later to the males with low levels of courtship and sooner to those with high levels of courtship. They would thus tend to mate preferentially with the sexually more active males.

As the numbers in the mating groups become smaller, so the difference between the P and R models becomes less. When the females arrive to mate singly, both P and R models produce results identical to those of the S models in which continuous mating is assumed, each female removing a male successively. Thus, when the weekly intervals in table 7.1 are reduced to single days, the P, R and S models all give the same estimates of the mating preference parameters. Table 7.2 gives the estimates of these parameters (O'Donald and Davis, 1977). (The details of the computer models have been published in O'Donald and Davis, 1977, who also gave the likelihood surfaces of the models.) In Model 4P, 13.7 per cent of the females prefer the dark males and 29.2 per cent prefer the intermediates. Davis and O'Donald (1976a) obtained strong evidence for assortative mating of intermediates: they estimated at maximum likelihood that 45.2 per cent of intermediates mate assortatively. Since intermediates form 60 per cent of the population, the assortative mating preferences for intermediates are exercised by 27.1 per cent of all females. This estimate is therefore in striking agreement with the independent estimate obtained from the breeding times. Since there is no assortment in the preference for dark males, however, different mechanisms may determine the female preferences for the two melanic phenotypes. This in turn suggests that Model 4P with separate mating preferences for the different genotypes is the most appropriate model of sexual selection for the Arctic Skua. If females exercise their preferences because they have a lower threshold towards the males they prefer, then preferential matings will usually take place before random matings, thus corresponding to the P models of partial preferences in which females make their choice from a group of males. In the P models of mating preference for the Arctic Skua, the group of males consists of all unmated males holding territories. Unlike the P models in polygynous mating systems, the preferred males who mate monogamously are removed from the group so that eventually no preferred males are left for the females to choose from. It is satisfactory that the most realistic model – Model 4P – should have the highest likelihood when fitted to the data on the distributions of breeding times, even though Models 3P and 4P, 3R and 4R do not differ significantly in their likelihoods at the maximum. More data are rapidly being obtained, and together with behavioural observations, these may decisively refute some of the other models as well as Models 2P and 2R which have already been refuted.

7.4   Selection as a result of mating preferences

In the models of sexual selection for polygynous species analysed in Chapters 4, 5 and 6 the males who mate preferentially are also available to mate at random. If α, β and γ are the proportions of females with preferences for genotypes AA, Aa and aa, as for example in model 4, then there is a stable equilibrium point when the allele A is at a frequency

$$p_* = (\alpha + \tfrac{1}{2}\beta)/(\alpha + \beta + \gamma)$$

In the models for monogamous species like the Arctic Skua, however, the progress of selection must be computed in successive generations in order to determine the approach to equilibrium. The genotypic frequencies of the pale, intermediate and dark males in a particular generation are used to determine their different distributions of breeding times according to the computer models of the mating preferences when given both the proportions of females that prefer the different phenotypes, as estimated from the data, and the proportions of females coming into breeding condition in the successive intervals of the breeding season. The sexual selective coefficients can then be calculated from the relationship between breeding time and breeding success. Table 7.3 shows this relationship in the Arctic Skua. The genotype frequencies in the next generation are thus calculated and hence the new distributions of breeding times which determine the new selective coefficients. This is repeated in each generation until equilibrium is reached.

Unless natural selection opposes the sexual selection, the allele A always spreads through populations in Models 3P and 3R to replace a. In both Models 4P and 4R, the equilibrium frequency is

Table 7.3   *Fledging success and breeding time of pairs of Arctic Skuas*

Weeks of the breeding season	Proportion of pairs breeding	Relative fledging success
x		$w = 1 - \dfrac{(0.47354 - x)^2}{24.254}$
0	0.09836	0.99075
1	0.44672	0.98857
2	0.23975	0.90393
3	0.13524	0.73683
4	0.07992	0.48726

The relative fledging success is given by the quadratic function shown which is a very close fit to the actual data.

given by the same formula as that for the model with polygyny. When the mating preferences are large, there are slight differences at equilibrium from p_*, caused by grouping the data on the breeding times into a few discrete intervals.

As we have already noted (section 7.2), the progress of sexual selection in monogamous birds is generally frequency-dependent. O'Donald (1973a, 1974) analysed this effect. If the dark allele A is dominant in its effect on mating preference, then selection will take place according to Model 1P or 1R. Table 7.4 shows the final

Table 7.4   *Final selective coefficients of the allele, a, at the point of being eliminated from the population*

Value of mating preference for intermediates	Final value of selective coefficient	
	Model 1P	Model 1R
β	s_∞	s_∞
0.1	0.00665	0.0192
0.2	0.0147	0.0430
0.3	0.0247	0.0719
0.4	0.0374	0.107
0.5	0.0540	0.148
0.6	0.0767	0.196
0.7	0.110	0.253
0.8	0.161	0.318
0.9	0.254	0.392
1.0	0.478	0.478

The initial selective coefficient when A is at a frequency of $p_0 = 0.01$ is $s_0 = 0.0601$.

selective coefficients of the allele a when it has nearly been eliminated from the population by the sexual selection for A. At the smaller values of β the selective coefficient declines from its initial value of 0.0601: the frequency-dependence is thus negative. At the larger values of β, the frequency-dependence is positive.

In Models 4P and 4R, the polymorphic equilibrium, p_*, is reached. It is stable. O'Donald (1976) calculated the progress of selection towards this equilibrium given the estimates of the mating preferences and the relative fledging success of the Arctic Skuas. The equilibrium frequency is always given by equation 4.2.2:

$$p_* = (\alpha + \tfrac{1}{2}\beta)/(\alpha + \beta + \gamma)$$

This is the equilibrium frequency in the 'C' models with complete preferences and polygyny. Table 7.5 shows the rate of approach to the equilibrium when the gene frequency starts at $p_0 = 0.01$. Table 7.6 shows the relative fitnesses during the progress towards equilibrium. Given the preferences estimated for Model 4P $(\alpha + \beta = 0.482)$, there is hardly any frequency-dependence in the values of fitness. But smaller values of $\alpha + \beta$ give negative frequency-dependence and larger values give positive frequency-dependence. The same equilibrium is reached whatever the initial frequency: the equation for p_* is

Table 7.5 *Gene frequencies of the allele, A, for dark phase in Models 4P and 4R*

	Gene frequency of allele A	
Generation	Model 4P	Model 4R
0	0.01	0.01
10	0.0160	0.0160
20	0.0270	0.0269
40	0.0762	0.0758
60	0.195	0.195
80	0.381	0.386
100	0.557	0.570
200	0.693	0.695
300	0.695	0.696
400	0.695	0.696
	$p_* = 0.695$	$p_* = 0.697$

p_* is the equilibrium given by $p_* = (\alpha + \tfrac{1}{2}\beta)/(\alpha + \beta + \gamma)$ which holds for both polygynous and monogamous mating.

Table 7.6   *Relative fitnesses of the genotypes during the progress of sexual selection*

	Relative fitnesses		
Generation	AA	Aa	aa
0	1.000	1.000	0.901
10	1.000	1.000	0.900
20	1.000	0.999	0.897
40	1.000	0.998	0.886
60	1.000	0.972	0.865
80	1.000	0.945	0.852
100	1.000	0.998	0.893
200	0.947	1.000	0.879
300	0.947	1.000	0.879
400	0.947	1.000	0.879

These selective coefficients were computed from the estimated values of the mating preferences in the Arctic Skua which determine the proportions of males breeding during the different weeks of the breeding season shown in table 7.1. Given the proportions of dark AA, intermediate Aa and pale aa males breeding in each week, the average fitnesses were then calculated from the relative fledging success in each week as shown in table 7.3. These average fitnesses then determine the genotypic frequencies in the following generation and hence their new distributions of breeding times and fitnesses. These values were thus used to calculate the values in table 7.5.

the same for both polygynous and monogamous mating systems indicating a globally stable equilibrium.

The general reasons for these changes in fitness during this evolution can be explained by the changes in the frequencies of the preferred males. At the start of selection $p_0 = 0.01$. There are hardly any dark males and only 2 per cent of the males are intermediate. Thus all the preferred males, both darks and intermediates, find their mates during the first week of the breeding season. They all have the same maximal fitness because breeding success is at its greatest in the first week. As the gene for dark increases in frequency, the intermediate males at first increase in frequency much more rapidly than the darks because the intermediates are the heterozygotes. Some of them are left unmated at the end of the first week. They mate in the second week with females who have a slightly lower fledging success. As the gene continues to increase in frequency, the intermediates become the commonest phenotype while the darks still remain relatively rare. The darks thus remain the fittest males being the earliest to breed. Meanwhile, the intermediates lose some of their advantage over the pales having become more common than the females who prefer them. As equilibrium is approached, the frequency of the darks also exceeds the frequency of the females who prefer them and they too start to lose some of their advantage. At equilibrium, there is always a greater excess of darks over the preference for dark than there is of intermediates over the preference for intermediate. Thus on average the intermediate males find mates before the darks and hence have a higher fitness. This would give rise to heterozygous advantage and would therefore maintain the polymorphism. However, the actual frequency of the dark allele is 0.56 in the males and less than the computed equilibrium frequency. At this frequency, the fitnesses of the dark and intermediate males are almost exactly equal (see tables 7.5 and 7.6). In spite of their sexual selective advantage over the pales, the dark and intermediate birds are often a year older than the pales when they breed for the first time. This is sufficient to offset their sexual advantage (O'Donald and Davis, 1976) and prevent a further increase of the dark allele.

7.5   Territory size and sexual selection

Territorial behaviour can be observed in many animals, especially the higher vertebrates. Territories are often assumed to be places where food is collected. However, many birds defend large territories although they do not feed within them. Crook (1962, 1965), in his study of pair formation and social organization of weaver birds, suggested that a large territory may help to conceal the nest. This would explain the different sizes of territories of the different species of *Euplectes*: species which nest in taller and denser vegetation have smaller territories (Emlen, 1957).

Territory may also have selective advantage as an area for sexual display and breeding. This is to be expected if territories are used exclusively for display and breeding and give no protection from predators. Wilson (1975) has listed examples from a large number of studies in which the functions of territoriality have been deduced from activities that take place within the territories: territories may be feeding areas, areas for sexual display and mating, or areas for shelter and concealment. It is argued that territoriality gains a selective advantage from one or more of these functions, but no direct evidence of the selective advantage has been obtained. The argument is always inductive based on comparisons between species. Such arguments, which were the only arguments Darwin himself could use, are very convincing when a particular character like territoriality differs between species, genera, or sub-families in relation to differences in ecology. But an argument that depends on an association between territoriality and other forms of social behaviour is less convincing. For example the association between feeding behaviour and territoriality does not logically require that feeding behaviour evolved first: territoriality might have evolved first and thus be the cause of the evolution of specific feeding behaviour.

Davis and O'Donald (1976b) and O'Donald (1977b) suggested that a larger territory may often be of direct selective advantage to a defending male: it will increase his chances of meeting a female. Territoriality will thus evolve by sexual selection. Davis and O'Donald (1976b) obtained data on territoriality in the Arctic Skua. From these data, the selective advantage of territory size can be calculated. Davis and O'Donald found a significant correlation between territory size and breeding time among those males with previous breeding experience who were taking a new mate. Males with larger territories breed earlier in the season. This effect is reduced, and is not quite significant statistically, if the males are breeding for the first time. There is no correlation at all in pairs which have bred together in previous years. Figure 7.1 shows the data of territory size and breeding time when experienced males mate with a new female. Since earlier breeding is correlated with higher reproductive success, larger territories are directly favoured by sexual selection. O'Donald (1977b) pointed out that this selection might itself be a consequence of the sexual selection for melanic males, since melanic males have larger territories and breed earlier in the season as figure 7.1 shows. The difference in territory size between melanic and non-melanic males is not, however, statistically significant given these rather small sample sizes, although the overall correlation with breeding date is highly significant. If territory size and mating success are both determined by levels of testosterone, as shown by Watson (1970) in red grouse; the earlier breeding and greater reproductive success of the males with

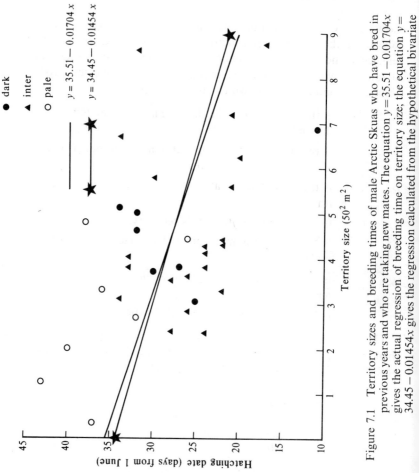

Figure 7.1 Territory sizes and breeding times of male Arctic Skuas who have bred in previous years and who are taking new mates. The equation $y = 35.51 - 0.01704x$ gives the actual regression of breeding time on territory size; the equation $y = 34.45 - 0.01454x$ gives the regression calculated from the hypothetical bivariate

larger territories would be a consequence of their increased mating success.

As an alternative hypothesis, size of territory might be the prime cause of mating success. Suppose the females arrive and land at random on the males' territories. They will be more likely to land on the larger territories. Males with larger territories will thus find a mate sooner than those with smaller territories: they will breed earlier on average in the breeding season and gain selective advantage from the greater reproductive success of earlier pairs. The females in the model are assumed to arrive in succession and land at random within the breeding grounds. To determine the probabilities of mating, the following symbols will be used.

p_{ij}	is the probability that the jth male is unmated when the ith female arrives ($p_{1j}=1$).
$p_{i+1,j}$	is the probability that the jth male is unmated when the $(i+1)$th female arrives.
P_{ij}	is the probability that the jth male mates with the ith female.
x_j	is the size of the jth male's territory.
w_j	is the average number of chicks fledged by the jth male.
W_i	is the average number of chicks fledged by the ith female.
b_i	is the breeding date of the ith female.

The probability that the ith female mates with the jth male is then assumed to be proportional to the size of the territory and the probability that the male is still unmated. Thus

$$P_{ij} = p_{ij}x_j \left/ \sum_j [p_{ij}x_j] \right.$$

If $\quad P_{ij} < p_{ij}$ then $p_{i+1,j} = p_{ij} - P_{ij}$

If $\quad P_{ij} \geq p_{ij}$ then $P_{ij} = p_{ij}$ and $p_{i+1,j} = 0$

$$w_j = \sum_i [P_{ij}W_i]$$

The values of W_i have been taken from the values in table 7.3 for the Arctic Skua. But these values can be altered to test the effect of different relationships between breeding time and reproductive success. For example O'Donald (1977b) also investigated the case when the greatest reproductive success was obtained by birds breeding at the mean, rather than at the beginning, of the breeding season. The advantage of larger territories is then much reduced, but not completely lost, since the males with the smallest territories always breed last with low reproductive success.

For the Arctic Skua the data of figure 7.1 can be used to calculate the probability P_{ij} that the jth male mates with the ith female. There are thirty-five males and females whose territory sizes and breeding times are shown in the figure. According to the model, the correlation coefficient of territory size and breeding time should be -0.3941: the actual coefficient is -0.4622, which is not significantly different from the hypothetical value. The regression according to the model should be $y = 34.45 - 0.1454x$. The actual and hypothetical regressions are plotted on the figure to show how closely they agree. But we know that female Arctic Skuas probably do not arrive and land at random on the territories as the model assumes: the females have often come from nearby territories where they mated in previous years. No doubt there must be some random element in the females' arrivals and hence some advantage gained by males with larger territories as a result of the increased chance of mating that their larger territories bring. But in the Arctic Skua this may be only a small component of the males' overall selective advantage. Their mating success is more likely to be determined by some other cause, such as hormone level, which pleiotropically increases territory size.

In general, however, if female arrivals were random, then territory size would certainly be selected as O'Donald's (1977b) computer simulations showed. At the same time, genetic factors, such as melanism that may have pleiotropic effects on territory size, would also be selected. If melanic male Arctic Skuas do have larger territories, this would be an hypothetical explanation of the sexual selection for melanism. From the bivariate frequency density P_{ij}, calculated from territory sizes of the melanic and non-melanic Arctic Skuas, the mean breeding dates of the males can then be calculated from their territory sizes shown in figure 7.1. These calculations are given in table 7.7. But, alas, in spite of the excellent agreement of hypothetical mean breeding date with observed date, territory size is not likely to be the main cause of sexual selection

Table 7.7   *Comparison of observed breeding dates with hypothetical breeding dates if a male's chance of mating depends on his territory size*

	Mean breeding date (measured in days from 1 June)	
Male phenotype	Observed	Hypothetical
Dark	29.22	29.70
Intermediate	30.05	29.84
Pale	34.09	33.78

for melanism in the Arctic Skua. Sexual selection takes place mainly in groups of young birds who may not be holding territories before they breed for the first time. Variation in territory size remains, however, a theoretically plausible mechanism for mate selection in many monogamous species of birds and other vertebrates. If territory size were normally distributed with a mean and variance equal to the mean and variance in the territory sizes of the Arctic Skua, then by taking a large sample from such normally distributed values the bivariate distribution of territory size and breeding time can be computed and hence the reproductive success of the males in an hypothetical population. This is shown in figure 7.2 for two cases: (i) when the optimum breeding time is at the beginning of the breeding season; (ii) when the optimum breeding time is at the mean breeding date. When the optimum breeding date is the mean, it is then just as disadvantageous to breed early in the season as it is to breed late. We may expect that males with the largest territories, who therefore have the greatest chance of breeding early, would suffer a selective disadvantage. But analysis of the probabilities of mating shows that the males with the largest territories are only slightly more likely to mate early: at the start of the breeding season all the males are unmated; the largest territory is only a small proportion of the total breeding area; and the male with the largest territory has only a slightly increased chance of mating at the beginning of the season. The males with the smallest territories, however, have a negligible chance of mating until most of the other males have mated: they are almost always left to mate with the last females to breed and always suffer a great selective disadvantage. Selection still tends to increase territory size. In figure 7.2, when the optimum breeding date is the mean, the fitness

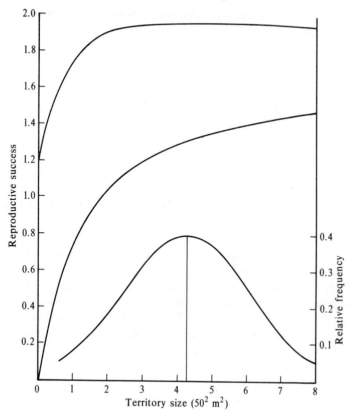

Figure 7.2 Fitness functions of territory size. The lower curve shows the
normal distribution of territory size. The middle curve shows the
the fitness function when the optimum breeding date is the
beginning of the season: fitness increases over the whole range of
territory sizes. The upper curve shows the fitness function when
the optimum breeding date is the mean: fitness only increases at
the smaller territory sizes. The difference in the average fitness in
the two cases is arbitrary and caused by the difference in the
proportion of pairs breeding at about the optimum date.

function of territory size decreases only slightly at the largest values
of territory size, but drops sharply at the smallest values. The
overall effect of selection will be to raise the mean territory size.
When the optimum breeding date is the beginning of the season,
then fitness increases rapidly with increasing territory size giving
very strong selection in favour of larger territories.

7.6   Mathematical analysis of monogamous mating systems

In the models analysed in the previous sections of this chapter, sexual selection has been assumed to take place by the mechanism that Darwin originally postulated – that the later matings have a lowered fertility or reproductive success. Then, since preferential matings tend to occur before random matings, the preferred phenotypes gain an advantage as a result of their higher average fertility. In the computer models of section 7.4, we saw that exactly the same equilibria are reached as in models with polygynous matings. Thus if α, β and γ are the separate preferences for AA, Aa and aa genotypes, then the allele A reaches equilibrium at a frequency

$$p_* = (\alpha + \tfrac{1}{2}\beta)/(\alpha + \beta + \gamma)$$

as in the simple polygyny models with complete expression of the preferences (described and analysed in section 4.2). This result holds for monogamy with a variety of different functions of reproductive success in relation to breeding time including the empirical function observed in the Arctic Skua (O'Donald, 1974). Similarly, if α and γ are preferences for dominant and recessive phenotypes, A and a, an equilibrium is always reached at a frequency of a given by

$$w_* = \gamma/(\alpha + \gamma)$$

I now analyse a simplified model of sexual selection with monogamous mating in which it can be shown mathematically that these results hold for any reduction in the fertility of later matings. The model has been simplified in comparison with the computer models of section 7.4 by making the assumption that all preferential matings take place before any of the random matings. The random matings then suffer from a reduction of fertility or reproductive success. As in the computer models, if a female with a preference cannot find an unmated male of the type she prefers, she then mates at random later, at the same time as the females with no preferences. The computer models are more complicated because preferential and random matings can both take place in each of a number of successive periods into which the breeding season is divided: males who are left unmated in one period may be chosen as mates by females coming into breeding condition in the next period. In the simplified models of this section, there are just two periods: preferential matings take place in the first and random matings in the second.

Two phenotypes with dominance

The phenotypes A (genotypes AA and Aa at frequencies u and v) and a (genotype aa) are preferred by proportions α and γ of the females. Since, in monogamy, males who have found a mate do not mate again (they are sampled 'without replacement' by the females!), there may be too few of one type of male to satisfy the numbers of females that prefer them. Three cases must be analysed: case (i), in which $\alpha \leq u + v$, $\gamma \leq w$; case (ii), in which $\alpha \leq u + v$, $\gamma > w$; and case (iii), in which $\alpha > u + v$, $\gamma \leq w$. The conditions which define these different cases will change as the frequencies change: if one phenotype is rare initially, the conditions for either case (ii) or (iii) will hold, but as the rare phenotype increases in frequency, the conditions may change into those of case (i).

Assume the conditions for case (i) hold: males with phenotypes A and a mate preferentially with frequencies α and γ, and randomly with frequencies $u + v - \alpha$ and $w - \gamma$. Among the A phenotypes, the genotypes AA and Aa may be chosen at random. Therefore AA males mate at random at the frequency

$$(u + v - \alpha) u / (u + v) = u(1 - \alpha - w)/(1 - w)$$

Similarly Aa males mate at random at the frequency $v(1 - \alpha - w)/(1 - w)$; and aa males at the frequency $w - \gamma$. Females mate at random with genotypes AA, Aa and aa at frequencies $u(1 - \alpha - \gamma)$, $v(1 - \alpha - \gamma)$ and $w(1 - \alpha - \gamma)$. The total frequency in random matings of both sexes is $1 - \alpha - \gamma$. Random matings have reduced fertility compared to preferential matings: the ratio of fertilities of random and preferential matings is $f : 1$ ($1 \geq f \geq 0$). Table 7.8 gives the relative fertilities of the different matings. When $f = 1$, the relative fertilities become simply the frequencies of random mating; the genotypes follow the Hardy-Weinberg Law. Thus, as we should expect, sexual selection takes place only if $f < 1$. The following recurrence equations give the genotypic frequencies in one generation and the next:

$$Tu' = p^2 - p^2(1 - f)(1 - \alpha - w)/(1 - w)$$

$$Tv' = 2pq - p(1 - f)(2q - w)(1 - \alpha - w)/(1 - w)$$
$$\quad\quad - p(1 - f)(w - \gamma)$$

$$Tw' = q^2 - q(1 - f)(q - w)(1 - \alpha - w)/(1 - w) - q(1 - f)(w - \gamma)$$

Hence we obtain the recurrence equation for the gene frequency of

Table 7.8. *Mean fertilities of monogamous matings with preferences for dominant and recessive phenotypes*

Mating type	Mean fertility
$AA \times AA$	$[\alpha u^2/(1-w) + fu^2(1-\alpha-w)/(1-w)]/T$
	$= [u^2 - u^2(1-f)(1-\alpha-w)/(1-w)]/T$
$AA \times Aa$	$[2\alpha uv/(1-w) + 2fuv(1-\alpha-w)/(1-w)]/T$
	$= [2uv - 2uv(1-f)(1-\alpha-w)/(1-w)]/T$
$AA \times aa$	$[\alpha uw/(1-w) + \gamma u + fuw(1-\alpha-w)/(1-w) + fu(w-\gamma)]/T$
	$= [2uw - uw(1-f)(1-\alpha-w)/(1-w) - u(1-f)(w-\gamma)]/T$
$Aa \times Aa$	$[\alpha v^2/(1-w) + fv^2(1-\alpha-w)/(1-w)]/T$
	$= [v^2 - v^2(1-f)(1-\alpha-w)/(1-w)]/T$
$Aa \times aa$	$[\alpha vw/(1-w) + \gamma v + fvw(1-\alpha-w)/(1-w) + fv(w-\gamma)]/T$
	$= [2vw - vw(1-f)(1-\alpha-w)/(1-w) - v(1-f)(w-\gamma)]/T$
$aa \times aa$	$[\gamma w + fw(w-\gamma)]/T$
	$= [w^2 - w(1-f)(w-\gamma)]/T$

The mean fertility over all different mating types is $T = 1 - (1-f)(1-\alpha-\gamma)$.

the allele A:

$$Tp' = p - p(1-f)[(1-\tfrac{1}{2}w)(1-\alpha-w)/(1-w) + \tfrac{1}{2}w - \tfrac{1}{2}\gamma]$$

The change in gene frequency from one generation to the next is therefore

$$T\Delta p = \tfrac{1}{2}p(1-f)[w(\alpha+\gamma) - \gamma]/(1-w)$$

so that at equilibrium

$$w_* = \gamma/(\alpha+\gamma)$$

This is exactly the equilibrium reached in the polygyny model. As in all models of sexual selection with no assortment of the matings, the Hardy–Weinberg Law holds at equilibrium: $u_* = p_*^2$, $v_* = 2p_*q_*$, $w_* = q_*^2$

Thus, assuming at equilibrium that $w = q^2$ and $dw/dp = -2q$, we get

$$\left.\frac{d\Delta p}{dp}\right|_{p=p_*} = -\frac{p_*q_*(1-f)(\alpha+\gamma)^2}{\alpha[1-(1-f)(1-\alpha-\gamma)]}$$

This is negative for all values $\alpha, \gamma > 0$, $1 \geq f > 0$. The equilibrium is therefore stable.

The recurrence equations for p and w can be used to obtain the gradient matrix in the region of equilibrium and thus used to determine the rate of approach to the equilibrium. If we put $p = p_* + x$, $q = q_* - x$ and $w = w_* - y$, where x and y are small deviations from the equilibrium values of p and w, then we have

$$
\begin{bmatrix} x' \\ y' \end{bmatrix} = \begin{bmatrix} 1 & -\left\{ \dfrac{p_*(1-f)(\alpha+\gamma)^2}{2\alpha[1-(1-f)(1-\alpha-\gamma)]} \right\} \\ 2q_* & -\left\{ \dfrac{p_*q_*(1-f)(\alpha+\gamma)^2}{\alpha[1-(1-f)(1-\alpha-\gamma)]} \right\} \end{bmatrix} \begin{bmatrix} x \\ y \end{bmatrix}
$$

The largest eigenvalue

$$
\lambda = 1 - \frac{p_*q_*(1-f)(\alpha+\gamma)^2}{\alpha[1-(1-f)(1-\alpha-\gamma)]}
$$

is the multiplying factor giving the geometric rate of approach to the equilibrium frequency, p_*. It is identical to the quantity

$$
\lambda = 1 + \frac{\mathrm{d}\Delta p}{\mathrm{d}p}\bigg|_{p = p_*}
$$

These results may be compared to those for the corresponding polygyny model in which the later, random matings also suffer a corresponding reduction in fertility. Then from equations (4.1.2) of section 4.1, Chapter 4, we immediately obtain the recurrence equation

$$
Tp' = \alpha p(1 - \tfrac{1}{2}w)/(1-w) + \tfrac{1}{2}\gamma p + fp(1-\alpha-\gamma)
$$

and

$$
T\Delta p = \tfrac{1}{2}p[\alpha/(1-w) - (\alpha+\gamma)]
$$

with exactly the same equilibrium frequency as in the monogamy model ($f < 1$) and in the polygyny model with $f = 1$. The geometric rate of approach to the equilibrium is given by the factor

$$
\lambda = 1 - \frac{p_*q_*(\alpha+\gamma)^2}{\alpha[1-(1-f)(1-\alpha-\gamma)]}
$$

These results are summarized in table 7.9. The only difference between sexual selection in polygynous and monogamous species lies in the difference between the rates of approach to the same equilibrium frequency. Monogamy is a much slower process unless the random matings have drastically reduced fertility. Not until $f = 0$ does the rate of approach become the same in the two mating

Table 7.9. *Equilibrium frequencies and rates of approach to equilibrium in models with either polygynous or monogamous mating*

Model	Equilibrium frequency	Geometric rate of approach to equilibrium
Equal fertility for all matings	Monogamy $w_* = q_0^2$	$\lambda = 1$
	Polygyny $w_* = \gamma/(\alpha + \gamma)$	$\lambda = 1 - p_* q_* (\alpha + \gamma)^2 / \alpha$
Fertility reduced in later matings	Monogamy $w_* = \gamma/(\alpha + \gamma)$	$\lambda = 1 - \dfrac{p_* q_* (1 - f)(\alpha + \gamma)^2}{\alpha [1 - (1 - f)(1 - \alpha - \gamma)]}$
	Polygyny $w_* = \gamma/(\alpha + \gamma)$	$\lambda = 1 - \dfrac{p_* q_* (\alpha + \gamma)^2}{\alpha [1 - (1 - f)(1 - \alpha - \gamma)]}$

In this table q_0 is the initial frequency of the allele a. With monogamy and equal fertility of all matings, the Hardy–Weinberg Law holds for all generations following the first.

systems. It is interesting that provided there is at least some reduction, however slight, in fertility, then monogamy always gives the same equilibrium as polygyny. Only at the point $f = 1$ does simple random mating replace sexual selection. If random matings were more fertile than preferential matings ($f > 1$), then in the model with monogamy we should have $\lambda > 1$: the same equilibrium would still exist but would not be stable. It would remain stable in the model with polygyny, however.

The foregoing analysis applies only when the conditions of case (i) hold. If they do not, then the conditions of either case (ii) or case (iii) must hold. Consider case (ii): since $\gamma > w$, not all the females who prefer a phenotypes can mate preferentially. The random matings take place with males and females at the following frequencies:

Genotypes	Males	Females
AA	$u(1 - \alpha - w)/(1 - w)$	$u(1 - \alpha - w)$
Aa	$v(1 - \alpha - w)/(1 - w)$	$v(1 - \alpha - w)$
aa	0	$w(1 - \alpha - w)$

Preferential and random mating then combine to produce the recurrence equation

$$Tp' = p - p(1 - f)(1 - \tfrac{1}{2}w)(1 - \alpha - w)/(1 - w)$$

where $T = 1 - (1 - f)(1 - \alpha - w)$. Thus according to this equation, at equilibrium we should have $w_* = 1 - \alpha$. But this is impossible because $\gamma > w$ and therefore $\gamma > 1 - \alpha$ or $\alpha + \gamma > 1$. The change in gene frequency

$$T\Delta p = -\tfrac{1}{2}pw(1 - f)(1 - \alpha - w)/(1 - w)$$

shows that if the allele a has just been introduced into the population at low frequency, so that $w \cong 0$, then

$$T\Delta p = -\tfrac{1}{2}pw(1 - f)(1 - \alpha)$$

$\Delta p < 0$, $p' < p$ and a increases in frequency. The state of fixation given by $(u, v, w) = (1, 0, 0)$ is therefore unstable. To find the stability of the other fixation state $(0, 0, 1)$, we put $1 - w = u + v$, $u \cong 0$, $v \cong 2p$ and $w \cong 1 - 2p$ to obtain

$$T\Delta p = \tfrac{1}{4}\alpha(1 - f) - \tfrac{1}{2}p(1 - f)(1 + \alpha)$$

For small p, $\Delta p > 0$ and $(0, 0, 1)$ is unstable.

Similar results are found for case (iii) when $\alpha > u + v$ and $\gamma \leq w$. The recurrence equation is then

$$Tp' = p - \tfrac{1}{2}p(1 - f)(w - \gamma)$$

where

$$T = 1 - (1 - f)(w - \gamma)$$

Therefore

$$T\Delta p = \tfrac{1}{2}p(1 - f)(w - \gamma)$$

showing that when $w \cong 1$, $\Delta p > 0$, and when $w \cong 0$, $\Delta p < 0$. Both fixation states are thus unstable, while the equilibrium $w_* = \gamma$ is impossible. These analyses of cases (ii) and (iii) suggest that changes in genotypic frequencies ultimately give rise to the conditions and equilibrium of case (i), regardless of whether the initial conditions were those of case (i) or cases (ii) or (iii).

Separate preferences for each genotype

Proportions α, β and γ of the females prefer males with genotypes AA, Aa, and aa. There are now seven possible sets of initial conditions.

Case (i): $\alpha \leq u$, $\beta \leq v$, $\gamma \leq w$

Case (ii): $\alpha \leq u$, $\beta \leq v$, $\gamma > w$

Case (iii): $\alpha \leq u$, $\beta > v$, $\gamma \leq w$

Case (iv): $\alpha > u$, $\beta \leq v$, $\gamma \leq w$

Case (v): $\alpha \leq u$, $\beta > v$, $\gamma > w$

Case (vi): $\alpha > u$, $\beta \leq v$, $\gamma > w$

Case (vii): $\alpha > u$, $\beta > v$, $\gamma \leq w$

The matings for case (i) have fertilities as shown in table 7.10. As in the model with dominance, when later matings are as fertile as earlier matings ($f = 1$), then the values in the table become simply the frequencies of random mating, the Hardy–Weinberg Law holds from the outset, and no sexual selection takes place. When later matings are less fertile ($f < 1$), we obtain the recurrence equations

$$Tu' = fp^2 + p(1-f)(\alpha + \tfrac{1}{2}\beta)$$

$$Tv' = 2fpq + (1-f)(\alpha q + \tfrac{1}{2}\beta + \gamma p)$$

$$Tw' = fq^2 + q(1-f)(\gamma + \tfrac{1}{2}\beta)$$

giving the linear recurrence equation for p

$$p' = p\left\{\frac{f + \tfrac{1}{2}\theta(1-f)}{f + \theta(1-f)}\right\} + \frac{\tfrac{1}{2}\phi(1-f)}{f + \theta(1-f)}$$

where $\theta = \alpha + \beta + \gamma$ and $\phi = \alpha + \tfrac{1}{2}\beta$ as in the models of complete preferences for each genotype analysed in section 4.2, Chapter 4. The recurrence equation has the complete solution

$$p_{(n)} = p_* + (p_0 - p_*)\left\{\frac{f + \tfrac{1}{2}\theta(1-f)}{f + \theta(1-f)}\right\}^n$$

Table 7.10. *Mean fertilities of monogamous matings with separate preferences for each genotype*

Mating type	Mean fertility
$AA \times AA$	$[\alpha u + fu(u - \alpha)]/T$
$AA \times Aa$	$[\alpha v + \beta u + fv(u - \alpha) + fu(v - \beta)]/T$
$AA \times aa$	$[\alpha w + \gamma u + fw(u - \alpha) + fu(w - \gamma)]/T$
$Aa \times Aa$	$[\beta v + fv(v - \beta)]/T$
$Aa \times aa$	$[\beta w + \gamma v + fw(v - \beta) + fv(w - \gamma)]/T$
$aa \times aa$	$[\gamma w + fw(w - \gamma)]/T$
	$T = 1 - (1 - f)(1 - \alpha - \beta - \gamma)$

showing that there is global convergence to the equilibrium $p_* = \phi/\theta$ at the geometric rate

$$\lambda = \frac{f + \frac{1}{2}\theta(1-f)}{f + \theta(1-f)} = \frac{1-(1-f)(1-\frac{1}{2}\theta)}{1-(1-f)(1-\theta)}$$

In a polygynous mating system, convergence to the same equilibrium takes place at the rate

$$\lambda = 1 - \tfrac{1}{2}\theta$$

If we equate the values of λ for monogamous and polygynous mating, we find that when $f = (1-\theta)/(2-\theta)$ the rates of sexual selection are equal. Thus for small values of θ, we would have to have $f \cong \frac{1}{2}$ to give the same rate of convergence as with polygyny; for a value $\theta = \frac{1}{2}$, then $f = \frac{1}{3}$ would give the same rate of convergence.

Some of the other cases of the model can easily be shown to lead to an impossible equilibrium frequency. Case (v), for example, in which $\alpha \leq u$, $\beta > v$, $\gamma > w$, gives an equilibrium at which

$$p_* = (\alpha + \tfrac{1}{2}v_*)/(\alpha + v_* + w_*)$$

$$q_* = q_*/(\alpha + v_* + w_*)$$

and hence at which

$$\alpha + v_* + w_* = 1$$

$$\alpha = u_*$$

But according to the conditions of case (v) we must have

$$\beta + \gamma > v_* + w_* = 1 - u_*$$

which entails

$$\alpha + \beta + \gamma > 1$$

an impossibility. A similar impossibility is found at the equilibrium of case (vii). However, for case (vi), the recurrence equation

$$Tp' = p - \tfrac{1}{2}(1-f)(v-\beta)(p+\tfrac{1}{2})$$

gives the difference equation

$$T\Delta p = \tfrac{1}{2}(1-f)(v-\beta)(p-\tfrac{1}{2})$$

The equilibrium $v_* = \beta$ is an impossibility since it entails $\alpha + \beta + \gamma > 1$ The equilibrium $p_* = \frac{1}{2}$ gives the genotypic frequencies

$u_* = \frac{1}{4}$, $v_* = \frac{1}{2}$ and $w_* = \frac{1}{4}$ which are possible according to the conditions of case (vi) but which do not represent a stable equilibrium since

$$\left. \frac{\mathrm{d}\Delta p}{\mathrm{d}p} \right|_{p=1/2} = \frac{\frac{1}{2}(1-f)(\frac{1}{2}-\beta)}{1-(1-f)(\frac{1}{2}-\beta)}$$

The equilibrium would only be stable if $\beta > \frac{1}{2}$ which would then contradict the condition $\beta \leq v$. Convergence to one of the fixation states will thus take place. If $\beta = 0$, then obviously $p < \frac{1}{2}$ represents the domain of convergence to (0, 0, 1) and $p > \frac{1}{2}$ the domain of convergence to (1, 0, 0). However, we may expect case (i) to provide the most generally applicable solution. Sexual selection in a monogamous species will normally lead to the same equilibrium as sexual selection in polygynous species.

8

Evolution of mating preferences

8.1 Fisher's theory of the evolution of mating preferences

Mating preferences are the consequence of variation in female response to male courtship. If some males court the females more actively, they will more often or more rapidly obtain a response from the females; their success relative to the less active males will be greater with the females with higher thresholds of response who require more stimulation, and who therefore exercise more choice in favour of the more active males. Average levels of female response may also vary to favour some males rather than others: female mating preferences may evolve if some females are genetically disposed to respond at different thresholds to different male phenotypes. Fisher (1930) put forward a suggestion that mating preferences might evolve as a result of the selective advantage gained by females who mate with the fittest males. The males may be fitter because they produce fitter offspring for the female who chose them or because they are better parents in the defence and care of their offspring. If some females can discriminate between the males and choose the fitter ones to mate with, they will gain an advantage over the others who cannot discriminate or exercise a choice. Fisher did not consider the behavioural mechanism by which such discrimination or choice might be achieved but simply assumed that appropriate genetic variation would be available for selection to act upon. In a passage quoted in Chapter 1, section 1.2, he explained that since the sons of such matings will tend to carry the genetic factors that determine both the female preference and the increased fitness, the preference will be selected through the selective advantage of the sons. Since the preference itself increases this advantage, there occurs what Fisher called a 'runaway process' in which both the preferred male characteristic and the mating preference evolve together. The character by which the females discriminate between the males would thus continue to develop, according to Fisher, until the sexual selection was eventually balanced by natural selection: eventually, the character would be developed to a point at which it became deleterious to its possessors, the initial selective advantage having been lost. Fisher did not explain why the character would be selected towards such extreme development, for the initial mating preference can have evolved only to favour the character that provided the initial means of discrimination. The character could either be a pleiotropic effect of the genotypes for increased male fitness, or be closely linked to the loci for fitness. The further development of this character into the elaborately ornamented plumage for sexual display in birds like the peacock and pheasant may be explained as the result of a supernormal stimulus (Tinbergen, 1948) to which a greater responsiveness is shown than to the normal stimulus: females with preferences for the normal development of the character would show a greater preference towards more extreme developments. This is a well known behavioural phenomenon which can be explained by a shift in the position of maximum response after training in discrimination (Hanson, 1959). If animals are rewarded when they respond to one stimulus, and punished or not rewarded if they respond to another, their peak response after training will be displaced beyond the point at which they received a reward and in a direction away from the negative stimulus. This 'peak shift' may be a consequence of the asymmetry in the conditions for being rewarded. Staddon (1975) suggested that since in the evolution of behaviour a reward might correspond to favourable selection then the greater response to supernormal stimuli may be caused by asymmetrical selection pressures: if not to respond to a stimulus produces an additional cost in terms of a reduction in the chances of survival or the numbers of offspring while to respond to a supernormal stimulus has little or no extra cost, then this asymmetry in the fitness function of response to stimuli will produce the extra responsiveness to supernormal stimuli. If female mating preferences evolve as a result of an advantage in being more responsive to males with the advantageous character, the evolution of a peak shift in their responses would select for the more extreme developments of the character in the way Fisher postulated, and eventually produce a character so extreme that its mating advantage was balanced by its deleterious effects on survival. When this point is reached, the selection of the females' responses will no longer be asymmetrical: stabilizing selection will maintain the character at this point in its phenotypic expression.

Fisher also hinted at another possibility: it will become advantageous for the less fit males to mimic the fitter males by possessing the character that discriminates them, for after mating preferences have evolved, the mating advantage of the preferred males may become the largest component of the selection in their favour. The selection of genetic modifiers increasing the expression of the preferred character may at the same time give rise to its partial expression in the less fit males who do not initially have it, since by hypothesis it is a pleiotropic effect of the genes for increased fitness or closely linked to them. Recombination may also give rise to less fit males who mimic the fitter by possessing their character. Thus the character may evolve in the less fit males. Mimicry of the fitter by the less fit may further reduce the selective advantage of possessing the character in addition to its reduced advantage as a supernormal stimulus. Eventually, therefore, as Fisher stated, the runaway process must stop.

O'Donald (1967) analysed Fisher's theory in detail by extensive computer simulations of the evolution of mating preferences. In these simulations, one locus determined the preferred character and another locus the female preference. The matings were polygynous and the females exercised a complete preference if they possessed a particular allele at a locus which might or might not be linked to the locus for the preferred character. The models showed that a recessive character would always be selected faster than a dominant: a much higher level of linkage disequilibrium can be maintained between the two loci since there is less tendency for the preference for a recessive to recombine with unfavourable male characters. The most rapid selection occurs when the preferences arise while the preferred character is still rare: the association between the preference and the character then becomes very close and the allele for the preference can reach high frequencies. If a dominant character is selected, the mating preference gene only slowly increases in frequency and remains at a low frequency at equilibrium. Natural selection against the preferred character has a greater effect in stopping the sexual selection of a dominant than in stopping a recessive. If natural selection opposes sexual selection, as Fisher suggested it must do eventually, then very high intensities of natural selection are needed to prevent the fixation of a recessive gene by sexual selection; but much lower intensities are sufficient to stop the sexual selection of a dominant. Stable polymorphisms are established whenever the natural selection is sufficient to prevent fixation of the preferred character. This is a consequence of the frequency-dependent mating advantage of the males in these models with polygyny. Given the equilibrium frequencies of the allele that determines the preference, the equilibrium frequencies of the allele for the character are then given by equations 4.3.7 and 4.3.8 of section 4.3. These results were obtained by computer simulations of the simple 'C' models with complete preference for a single character. In later sections of this chapter, I discuss results obtained from more general models.

8.2   Zahavi's theory of preference for handicapped males

Zahavi (1975) put forward an extraordinary, even preposterous, idea that directly contradicts Fisher's theory. He suggested that females would prefer to mate with 'handicapped' males, who, as a result of their handicap, would have been subjected to more intense selection before mating. The survivors of the handicap would thus be fitter than survivors whose fitness had not been tested by extra selection imposed upon them by a handicap. The peacock's tail must handicap a male trying to escape from a predator. The handicapped males who survived to mate must therefore have been fitter – fitter in the ordinary sense of physically fitter – than males who had not been encumbered by long conspicuous tail feathers. According to Zahavi, the females should therefore mate with the handicapped males. Logically we could conclude that the greater the deformity a male suffers from, the greater should be the female preference in his favour – an absurdity which immediately refutes Zahavi's theory. Unfortunately, although two papers (Maynard Smith, 1976; Davis and O'Donald, 1976c) have been published in refutation of the theory, it has been discussed widely, though not in print, by ethologists some of whom seem to have taken the idea seriously. It has also been partly responsible for general condemnation of the whole subject of 'sociobiology', of which Zahavi's theory is deemed to be part (Lewontin, 1977), for the inadequacies of models on which many sociobiological explanations have been based. A general examination of the theory and its implied premises therefore seems to be necessary.

Zahavi first discusses Fisher's theory. Having stated Fisher's theory, he says:

It is obvious that with only one measurement at hand an exaggeration of the character, beyond its significance as an indicator for quality, will be undetected by females. But under most circumstances females probably select males by more than one character. In fact males advertise to females by voice, colour, movement, form, etc. An exaggeration of only one of these characters, without correlation to the quality of the males should lose its effect by negative selection. Any female which continues to prefer such a character will end up with a worse mate than females which choose males by all other characters except the exaggerated one. Generally it should be assumed that a character in a male is attractive to a female because it helps the female to select the better male.

Zahavi ignores the point that in Fisher's theory, the mating preference increases the initial advantage of being able to choose a fitter male and therefore allows the continued evolution of sexually exaggerated characters to the point at which they have become disadvantageous when considered apart from the mating preference in their favour. The number of characters that may be selected seems to have no relevance at all to Fisher's theory. Certainly the fitter males must be distinguishable before mating preferences can evolve in their favour. Discrimination will be easier if more characters vary between the males. Phenotypes will naturally tend to be selected for a combination of characters. If mating preferences can more easily be expressed for a combination of characters, the pleiotropic effects of a gene on each character may therefore be increased by the selection of specific modifiers. Fisher's theory is in no way limited to the selection of mating preferences for single characters. Indeed any character may be broken down into a combination of characters as any single measurement may be broken down into separate parts. What constitutes a single character is a matter of definition that is completely arbitrary. Zahavi seems to contradict himself in saying that an exaggeration of only one character will lose its effect by negative selection, since the exaggerated character would constitute exactly the sort of handicap that would test the quality of the male who possessed it. Zahavi makes no specific criticism of Fisher's theory, although rejecting it for what seem to be irrelevant or contradictory reasons.

In discussing his own theory, Zahavi puts forward no specific genetic model of the evolution of mating preferences. On the contrary he writes as if he thinks that mating preferences will be found in all sexually reproducing organisms. Having rejected

Fisher's theory, he says: 'When the handicap caused by mate preference is considered as the key to the selection process rather than its byproduct it is reasonable to expect handicaps ... ' Thus he seems to be implying that mating preferences are the cause of the evolution of the handicaps, giving the impression that he thinks females will automatically have preferences for handicapped males. Yet he also talks of females with a preference for handicapped males gaining a benefit because their males have been subjected to, and have survived, a severe 'test of quality' as a result of the handicap, implying that such females gain a selective advantage as a result of the fitter offspring they produce. Their preferences would thus evolve to favour handicapped males. Zahavi is obviously trying to discuss an evolutionary process of some sort. But what is supposed to be evolving? Is it handicaps and deformities of the males, or female preferences to mitigate these burdens?

To exemplify this weird theory, Zahavi takes an example of sexual selection in birds: the males are colourful with exaggerated sexual displays involving their colourful plumage; the females are cryptic. He says that the females cannot withstand the extra predation resulting from having the colourful plumage. This he calls the 'accepted explanation'. He then continues:

The interesting question is why some species 'find it better' to be excited or intimidated by more colourful males? This will only make sense if colourful males are of better quality. I suggest that a mature, colourful male has already proved itself to be of better quality than one with cryptic plumage since it has withstood the extra predation risk involved in its plumage. Hence colourful plumage is a mark of quality, and it is for the advantage of the females to be attracted by colourful males.

Evidently Zahavi thinks that females with a preference for colourful males gain an advantage because they are mating with a group of males that has been subjected to more intense natural selection by predators. The 'quality' will presumably be their ability to avoid predation. Since the females are supposed to gain an advantage by their preference, we certainly gain an impression that the theory concerns the evolution of female preferences, although Zahavi rejects models which allow for the advantage conferred by the mating preference. He never considers the genetics of the characters nor discusses exactly what advantage the females might have gained.

If the effect of natural selection is considered by itself, the colourful males with elaborate characters for sexual display must certainly be disadvantageous and selected more intensely than the cryptic males. This is a premise of Fisher's theory: the adverse natural selection eventually brings his 'runaway process' of sexual selection to a halt. It has been shown to be true of male stickle-backs. In some populations, as described in Chapter 4, section 4.3, the males are polymorphic: some develop the typical red throat in the breeding season; others do not. Moodie (1972) found that red sticklebacks are attacked many times more often by trout than non-red ones. Experiments by Semler (1971) showed that female stickle-backs lay their eggs for preference in the nests of red males. This can lead to a balanced polymorphism maintained by sexual selec-tion for the red males and natural selection against them (O'Donald, 1973b). Thus there are two phenotypes – cryptic and colourful. For each phenotype, some combination of characters must be optimum. When individuals whose phenotypes are furthest from their optima have been eliminated, a group of survivors with higher fitness will remain to be chosen as mates by the females. The colourful males, having been selected more intensely by predators, will have had their fitness raised to a higher level than the cryptic males. Zahavi clearly supposes that females with a preference to mate with these fitter males will gain an advantage as a result of their mates' higher fitness to escape from predators. This advan-tage could only accrue to offspring of these matings: for the females to gain any advantage at all they must produce offspring fitter on average than the offspring of females who mated with the cryptic males. Zahavi seems to take this for granted. But it is not true. For one thing, a proportion of offspring from matings with colourful males will themselves be males with the colourful phenotype. The proportion of colourful offspring produced will be determined by the genetics of the colourful character and the gene frequencies. No matter by what amount fitness may have increased, the colourful individuals must always, by hypothesis, have a lower fitness than the others: if not, there would be no 'test of quality'. These colourful individuals then lower the average fitness of the progeny. Davis and O'Donald (1976c) showed that in all realistic conditions, the aver-age fitness of the offspring of females who mated with the colourful males must be less than the average fitness of the offspring of the other females who mated with the cryptic males. It is simply not true to say that females gain an advantage by mating with the more colourful males who have been subjected to more intense natural selection. The logic of the argument is demonstrably false.

But first we should consider the validity of the premises of Zahavi's theory. He himself neither states the premises of his theory, nor analyses its consequences. We can only deduce the premises, working backwards from what Zahavi assumes the consequences to be. It is clearly assumed by implication that the effects of selection in the colourful males is passed on to the offspring of the females who prefer the colourful males, raising the fitness of their offspring. In no other way could the females gain any advantage. This in turn assumes that directional selection for the optimum combination of characters in the colourful phenotype is still taking place and that these characters are heritable with additive genetic variation. It also assumes that the selection of the colourful males raises the fitness of the cryptic males and females who are produced as offspring of matings with the colourful males. Thus we reach, by deduction, the premises of Zahavi's theory. Both colourful and cryptic phenotypes must be at a mean fitness that is well below the optimum for whatever combination of characters is being selected. If this were not so and the phenotypes were at their optimum combination of characters, no increase in fitness could be passed to the offspring of matings with the colourful males and hence no advantage could accrue to females with preferences. In addition to this basic premise, it is assumed that the same combination of characters must be advantageous for each phenotype. Taking these two premises for granted, an advantage might then be gained by females who exercised their preferences for colourful males, provided the lower fitness of colourful offspring produced by the preferential matings did not lower the average fitness of the progeny by more than the amount by which selection within the colourful males had raised it. Calculations show, however, that on these implicit premises the average fitness of the progeny of the preferential matings must always be lowered rather than raised. Maynard Smith (1976) used a three-locus model analysed numerically on a computer to show that no selection of a gene for the mating preference can take place. The preference gene only increases in frequency if it starts closely linked in coupling with the gene that raises fitness by selection within the group of colourful males. But this increase in frequency is merely the well known 'hitch-hiking' effect, which must always occur when any gene is in coupling with one that increases fitness. It is quite unrealistic to start the selection in complete coupling: the preference gene might just as well be in repulsion as in coupling.

Davis and O'Donald (1976c) considered directional selection of a quantitative character within the colourful males and analysed the average effect on the offspring of matings with colourful and cryptic males. There is of course no effect if the heritability of the character is zero. In most cases the selection had to be impossibly intense to produce any advantage for the preferential matings: in all cases it had to be much too intense to be biologically realisitc.

Maynard Smith's and Davis and O'Donald's calculations show that even if the essential premises of Zahavi's theory were true females would not gain an advantage by mating with handicapped males: the logic of the theory is demonstrably false. The basic premises are false, too. In the theory, both colourful and cryptic phenotypes must have a similar optimum combination of characters so that selection to raise the fitness of the colourful phenotype also produces an increase in the fitness of all the offspring, both colourful and cryptic, of a mating with a colourful male. Without this assumption, matings with colourful males could never be advantageous. Yet it is most unlikely that both phenotypes would have similar optima in their expression. To avoid predation, the more easily seen, colourful individual would need speed and agility to escape: the cryptic individual would need to remain still and merge into his background. Necessarily, the colourful and cryptic phenotypes will have different optimum combinations of characters and will therefore be subjected to disruptive selection for those combinations appropriate to their different ecological niches. Thus in fact females who preferred to mate with colourful males would suffer two disadvantages: some of their offspring would have the disadvantageous colourful phenotype; their cryptic offspring would have reduced fitness by inheriting a combination of characters inappropriate to their cryptic phenotype. Thoday (1960) selected *Drosophila* for high and low bristle numbers with different marker genes on the chromosome carrying the bristle number loci. These experiments show what would happen under this mode of disruptive selection. In Thoday's selected lines, combinations of genes were maintained in coupling with the markers. A stable state of linkage disequilibrium was quickly reached at which no further divergence in bristle number could take place owing to recombination between markers and bristle number genes. At this point no selection in the parents can produce any increase in fitness in the offspring: all additive genetic variance in fitness has been lost,

only the dominance and interaction variance remaining. In all
genetic models, additive genetic variance in fitness is eventually
lost. When the combination of characters is as near to the optimum
as it can get, there can be no advantage in mating with a more
highly selected male since no benefit can be received in an increase
in the fitness of the offspring. An increase in fitness can only be
passed on to the offspring during the period in which directional
selection is taking the combination of characters towards the
optimum and some additive genetic variance in fitness still remains.
This period will be brief if selection is intense. In Zahavi's theory
the disadvantage of mating with colourful males must continually
increase as a smaller and smaller part of the increase in fitness is
passed on to the offspring, thus producing a greater and greater
overall disadvantage arising from the colourful individuals among
the offspring.

We must therefore reject Zahavi's theory. It does not explain the
evolution of mating preferences in favour of elaborate and colourful male characteristics. Fisher's theory is the only satisfactory
explanation to have been put forward. Mating preferences first
evolve for males with characters that indicate increased fitness and
not for handicapped males with reduced fitness. Since the mating
preference adds to the initial advantage of the preferred character,
preference and character evolve together. It is advantageous for the
preferred males to be as distinct as possible and therefore at the
same time for the females to recognize these distinct males and
choose them as mates. The distinguishing characters thus continue
to evolve as a result of the evolution of the mating preferences. A
'peak shift' in female response (see section 8.1) will also help
towards the evolution of more exaggerated sexual characteristics.
These eventually must become deleterious: the 'runaway process'
comes to a halt as natural selection balances the sexual selection.
The exaggerated character which thus evolves can only be considered a handicap by ignoring its sexual advantage and con-
centrating solely on its disadvantage to survival: it is not really a
handicap at all, but only appears to be a handicap when viewed
from one aspect of its overall effect on fitness.

What Zahavi seems to regard as the cause of the evolution of
mating preference is in Fisher's theory the effect of that evolution:
the selective advantage of a distinct and colourful sexual display
increases as the mating preference evolves producing a character

which eventually acquires deleterious effects and becomes a handicap to survival. It is impossible that a handicap by itself could be a cause of the evolution of the mating preference.

8.3 Models of the evolution of preferences for dominant and recessive characters

In the models I shall describe, a second locus determines the preferences for the genotypes, AA, Aa or aa, of the first locus. The females may prefer the three genotypes separately, or A may be dominant to a. I have assumed, as I think it is necessary to assume, that there is always dominance between the alleles for the preferences. Suppose that three alleles B', B and b determine the preferences, with a dominance relationship $B' > B > b$. Then if A is dominant to a, females who are $B'B'$, $B'B$ or $B'b$ all prefer to mate with males who are AA or Aa; females who are BB or Bb prefer to mate with males who are aa. Thus the allele B' determines the preference for the A phenotype and the allele B determines the preference for the a phenotype: b is a 'null' allele, the females who are bb mating completely at random among the males. The alleles B', B and b have gene frequencies p_B, q_B and r_B in the population. To make the model more general and realistic, I have assumed that only a proportion of the females who possess the genotypes for the preferences actually express them. If α_x and γ_x are the proportions of females with preferential genotypes who actually express their preferences, then the mating preferences, α and γ, for phenotypes A and a will be as follows:

$$\alpha = \alpha_x \cdot [\text{frequencies of genotypes } B'B', B'B \text{ and } B'b \text{ in females}]$$

$$\gamma = \gamma_x \cdot [\text{frequencies of genotypes } BB \text{ and } Bb \text{ in females}]$$

α_x and γ_x can be considered as values of the 'penetrance' of the alleles B' and B.

When females prefer the genotypes AA, Aa or aa separately, with no dominance of the alleles A or a, then the genetics of the preferences is assumed to be as follows:

$B'B'$, $B'B$ and $B'b$ ♀♀ prefer AA ♂♂

BB and Bb ♀♀ prefer Aa ♂♂

bb ♀♀ prefer aa ♂♂

The proportions of these females who express their preferences are
then α_x, β_x and γ_x in this model with no dominance: these values
represent the penetrance of the alleles B', B and b in the females. If
α_x, β_x or $\gamma_x = 0$, then one of the alleles becomes a 'null' allele with no
effect on mating preference. According to the general premise of the
model, aa is the original wild-type, but the allele A is spreading
through the population to become the new wild-type. The genotype
Aa is the first to appear in the population, followed by AA after A
has become more common. A preference for Aa could only increase
in frequency if it were at least semi-dominant to the preference for
aa. Thus I assume that B is dominant to b, and also that B' is
dominant to B. In the models to be described in this section and
section 8.4, the females, who express their preference are a constant
proportion of females possessing the preferential genotypes. But, of
course, these proportions might also vary according to the frequencies with which the females encounter the males they prefer, thus
giving rise to 'P' and 'PF' models with partial frequency-dependent
expression of the preferences.

Since there are two alleles at locus A and three alleles at locus B,
a total of twenty-one different genotypes can be produced when all
the different coupling and repulsion heterozygotes are included.
Table 8.1 shows the preferential matings that take place either with
or without dominance of the preferred characters. Within the
groups of preferential or random matings, the different genotypes
all mate at random. This is equivalent to a random union of the
gametes produced by individuals who mate within the groups.
Thus the gamete AB' is produced at a frequency

$$a_1 = \tfrac{1}{2}\alpha_x(2y_1 + y_2 + y_3 + y_4 + (1-r)y_5 + (1-r)y_6 + ry_9 + ry_{13})$$

by females who exercise their separate preference for AA males.
This gamete is also produced at a frequency

$$u_1 = \tfrac{1}{2}(2y_1 + y_2 + y_3)/(y_1 + y_2 + y_3 + y_7 + y_8 + y_{12})$$

among the preferred AA males. Therefore the preferential matings
in favour of AA males produce individuals who are AB'/AB' at a
frequency a_1u_1. To this frequency must be added the frequency of
AB'/AB' individuals produced by the matings that take place at
random with all males. This gives the total frequency, y'_1, of the
genotype AB'/AB' in the next generation. The frequencies of the
other genotypes in the next generation, $y'_2, \ldots, y'_{21}$, are calculated
in a similar way. Recurrence equations are thus obtained giving the

Table 8.1   *Preferential matings in genetic models of the mating preferences*

Genotypes and frequencies		Preference for dominant or recessive phenotypes		Preference for separate genotypes		
		A	a	AA	Aa	aa
$AB'AB'$	y_1	α_x	–	α_x	–	–
AB'/AB	y_2	α_x	–	α_x	–	–
AB'/Ab	y_3	α_x	–	α_x	–	–
AB'/aB'	y_4	α_x	–	α_x	–	–
AB'/aB	y_5	α_x	–	α_x	–	–
AB'/ab	y_6	α_x	–	α_x	–	–
AB/AB	y_7	–	γ_x	–	β_x	–
AB/Ab	y_8	–	γ_x	–	β_x	–
AB/aB'	y_9	α_x	–	α_x	–	–
AB/aB	y_{10}	–	γ_x	–	β_x	–
AB/ab	y_{11}	–	γ_x	–	β_x	–
Ab/Ab	y_{12}	–	–	–	–	γ_x
Ab/aB'	y_{13}	α_x	–	α_x	–	–
Ab/aB	y_{14}	–	γ_x	–	β_x	–
Ab/ab	y_{15}	–	–	–	–	γ_x
aB'/aB'	y_{16}	α_x	–	α_x	–	–
aB'/aB	y_{17}	α_x	–	α_x	–	–
aB'/ab	y_{18}	α_x	–	α_x	–	–
aB/aB	y_{19}	–	γ_x	–	β_x	–
aB/ab	y_{20}	–	γ_x	–	β_x	–
ab/ab	y_{21}	–	–	–	–	γ_x

frequencies of the 21 genotypes from one generation to the next. I have made no attempt to analyse the model mathematically, but have used the recurrence equations to compute numerical results for particular initial frequencies and values of the parameters α_x, β_x and γ_x. The calculations were always started with the alleles at the two loci in linkage equilibrium and the genotypes at each locus in the Hardy–Weinberg ratios. Thus at the start of the calculations, the mating preferences were neither associated nor dissociated with the preferred genotypes: the matings were not assortative. But since a mating preference allele tends to segregate with the allele for the preferred character, linkage disequilibrium is produced, and hence some assortative mating develops. I have measured this by the correlation coefficient between particular alleles and the percentage assortment in the matings for particular phenotypes or genotypes. For example, among preferential matings for AA genotypes, a proportion of matings of the type $AA \times AA$ will be assortative. We can calculate the expected frequency with no assortment of these matings by assuming the genotypes are in linkage equilibrium. The excess of the actual frequency of these matings over its expectation is then expressed as a percentage of all preferential matings in favour of the AA genotypes. I have called this the percentage assortment. It is a function of the mating preference and the linkage disequilibrium between the alleles at the two loci. Some residual linkage disequilibrium and assortative mating is often left after the population has reached a final equilibrium state. As we shall see, the level of linkage disequilibrium that develops can be critical in determining the outcome of selection.

Table 8.2 gives the results of computing the equilibrium frequencies and progress to equilibrium when preferences favour a dominant or recessive character. It was assumed that 50 per cent of. females who are genotypically $B'B'$, $B'B$ or $B'b$ express a preference for the A phenotype: the homozygous and heterozygous penetrance of the allele B' is therefore $\alpha_x = 0.5$. Females who are genotypically BB or Bb express their preference for the a phenotype with a penetrance that varies from $\gamma_x = 0.1$ to $\gamma_x = 1.0$. The equilibrium frequencies of the allele A are then shown for particular initial values of the gene frequencies. Generally the final equilibrium frequencies are higher when initial frequencies are lower. The actual frequencies in the progress to equilibrium show why a higher equilibrium frequency is reached from a lower initial frequency. The linkage disequilibrium and hence the association between the alleles for the preference and preferred character is greater at lower frequencies. This produces greater assortment during the selection for the preferred phenotypes. The preference itself is also selected to a more significant extent.

The final equilibrium reached is always given by equation 4.1.5

$$p_* = 1 - \sqrt{[\gamma/(\alpha+\gamma)]}$$

Thus in the case when $\alpha_x = 0.5$, $\gamma_x = 0.1$ with initially $p_A = 0.1$ and $p_B = 0.05$, then we have the following equilibrium frequencies of females with preferential genotypes:

$B'B'$, $B'B$ and $B'b$ occur at a total frequency of 0.13035;

BB and Bb occur at a total frequency of 0.16540.

Table 8.2 *Equilibrium frequencies and the progress to equilibrium when one locus determines a dominant or recessive character and another the females that prefer each phenotype*

(i) *Equilibrium frequencies of the allele, A, in males*

The allele A determines the dominant character and a the recessive character. The penetrance of the preference for A is $\alpha_x=0.5$ throughout. p_B is the frequency of B'

Penetrance of preference for aa males γ_x	Initial gene frequencies		
	$p_A=0.1$ $p_B=0.05$	$p_A=0.01$ $p_B=0.05$	$p_A=0.01$ $p_B=0.005$
0.1	0.5507	0.6270	0.6391
0.2	0.4143	0.5056	0.5230
0.3	0.3302	0.4263	0.4471
0.4	0.2717	0.3680	0.3911
0.5	0.2286	0.3227	0.3473
0.6	0.1956	0.2862	0.3116
0.7	0.1697	0.2561	0.2820
0.8	0.1489	0.2309	0.2568
0.9	0.1318	0.2095	0.2352
1.0	0.1177	0.1911	0.2165

This table shows that the equilibrium frequencies of the allele A depend on the initial frequencies of both the A allele and the alleles for the preferences. In each case, the allele B, which determines the preference for aa, started at twice the frequency of the allele, B', which determines the preference for A: $q_B=2p_B$ initially.

(ii) *Progress to equilibrium when each preferential genotype has a penetrance of* $\alpha_x=\gamma_x=0.5$

Generation	Initial frequency $p_A=0.1$			Initial frequency $p_A=0.01$		
	p_A	p_B	percentage assortment	p_A	p_B	percentage assortment
0	0.1	0.05	0	0.01	0.005	0
1	0.1059	0.0500	8.83	0.0112	0.0050	12.04
2	0.1114	0.0503	12.66	0.0123	0.0051	17.49
5	0.1265	0.0515	15.88	0.0159	0.0053	22.67
10	0.1476	0.0533	15.52	0.0222	0.0058	22.94
20	0.1780	0.0556	14.22	0.0355	0.0065	21.84
50	0.2165	0.0580	12.71	0.0769	0.0078	19.27
100	0.2275	0.0586	12.31	0.1396	0.0089	16.07
200	0.2286	0.0586	12.27	0.2282	0.0099	12.21
500	—	—	—	0.3251	0.0106	8.71
1000	—	—	—	0.3459	0.0108	8.05
2000	—	—	—	0.3473	0.0108	8.00

This shows that from lower initial frequencies, a greater association can develop between the loci: hence the percentage assortment is greater, and the preference is more rapidly selected, producing a higher final equilibrium frequency. The percentage assortment is the frequency, expressed as a percentage of all preferential matings for the phenotype A, by which those matings of the type $A \times A$ exceed an expected frequency calculated on the assumption that the preferred and preference genotypes associate at random in linkage equilibrium.

Therefore

$$\alpha = \alpha_x(0.13035) = 0.065175$$

$$\gamma = \gamma_x(0.16540) = 0.016540$$

$$p_* = 1 - \sqrt{(0.01654/0.081715)}$$

$$= 0.5501$$

This predicted equilibrium frequency is very close to the actual frequency reached by the A allele. The discrepancy arises because the frequencies had not actually reached equilibrium at the time when the computation was stopped. Further, very slow gene frequency changes were still taking place. The final predicted equilibrium must continue to change as the frequencies of the alleles for the mating preferences change. Near equilibrium, the levels of preference in any generation predict a frequency that is very close to the actual frequency. The actual frequency can only change very slowly because its rate of change is a function of the difference between the predicted and actual frequency; its slow rate of change may continue for many generations, stopping only when the predicted equilibrium frequency at last coincides with the actual frequency. Equilibrium has then finally been attained, but many thousands of generations may have been needed to reach it.

Recombination fraction seems to have an almost negligible effect on the outcome of selection. This surprised me at first since higher levels of linkage disequilibrium can usually be maintained with closer linkage. In sexual selection however, closer linkage reduces linkage disequilibrium. This effect is partly caused by starting the computations in linkage equilibrium: when the loci are closely linked, recombination only slowly produces the favourable combinations in which alleles for the mating preferences are linked in coupling with alleles for the preferred characters. Close linkage actually opposes the powerful effect of preferential mating in producing these favourable combinations. When the loci are loosely linked, the preferential matings produce the favourable combinations in the first few generations. As table 8.2 shows, the percentage assortment increases rapidly in the first five generations. The increase in percentage assortment is much slower when linkage is tight. Table 8.3 shows the final equilibrium frequencies which are reached with different recombination fractions. With the same initial frequencies, the final equilibrium frequencies are very similar.

Table 8.3   *Equilibrium frequencies reached with different recombination fractions*

Recombination fraction	Initial frequency of the allele for the mating preference for A phenotypes	
	$p_B = 0.05$	$p_B = 0.005$
0.5	0.2286	0.2192
0.2	0.2259	0.2188
0.05	0.2190	0.2172

In each case, the allele A started at a frequency of $p_A = 0.1$. The penetrance was 0.5 for all preferential genotypes: $\alpha_x = \gamma_x = 0.5$.

In fact, although the degree of association between the loci (and hence the percentage assortment) is slower to build up when linkage is tight, very similar values are shown at equilibrium: the slower selection with tight linkage at the start is to some extent offset by more rapid selection later, producing very similar final equilibrium frequencies regardless of the amount of recombination.

The initial frequencies are of much greater significance in determining the rate of selection and the final equilibrium reached. When both the preferred character and the preference start at low frequencies, a high degree of association can be produced between the corresponding alleles at the two loci. This produces more assortative mating, more rapid selection and a higher final equilibrium frequency. Figure 8.1 shows the rates of selection with equal penetrance of the preferential genotypes ($\alpha_x = \gamma_x = 0.5$), but with very different initial gene frequencies. When $p_A = 0.001$ and $p_B = 0.0005$, a much higher final frequency is reached than when $p_A = 0.1$ and $p_B = 0.05$. In the course of selection starting from lower frequencies, higher levels of association are produced between the alleles A and B'.

In all these simulations of the evolution of mating preferences, I assumed that both phenotypes A and a will be the object of female preference: females with the allele B' prefer A; females with the allele B prefer a; and females who are bb and carry neither B' nor B mate completely at random. The levels of preference that then evolve predict the equilibrium frequencies of the alleles A and a in accordance with equation 4.1.5 derived in section 4.1 for the single locus model with fixed levels of the mating preferences. The equilib-

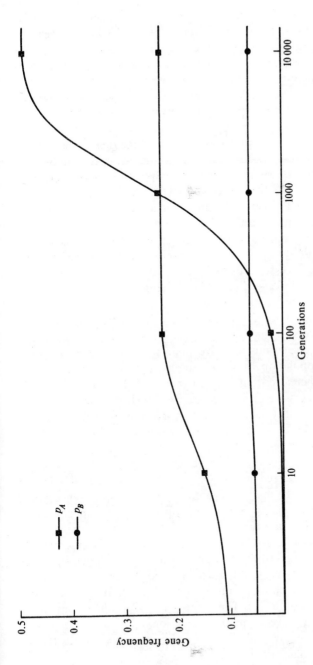

Figure 8.1 Rates of selection of a dominant gene with equal penetrance of the preferred genotypes but with different initial gene frequencies. In the computer model, 50 per cent of females who carry the allele B' express a preference for A as a dominant and 50 per cent of females who carry B (B' dominant to B) express a preference for a as a recessive. The model was run with two sets of initial gene frequencies: (i) $p_A^{(0)} = 0.1, p_B^{(0)} = 0.05$; (ii) $p_A^{(0)} = 0.001, p_B^{(0)} = 0.0005$. Values of p_B at the lower initial frequency are always small and are not shown in the figure.

rium frequencies in the 'C' models are all stable and poly-morphisms are therefore maintained by a balance between the relative mating advantage of the two phenotypes A and a. The stability of the equilibrium is of course a function of the frequency-dependent mating advantage of the two male phenotypes. Alter-natively, polymorphisms can be maintained if a mating preference evolves for one phenotype and is eventually balanced by natural selection in opposition to it. The mating preference can evolve to high levels if females prefer only one of the male phenotypes: very strong natural selection may then be necessary to stop the pre-ferred character spreading through the population to complete fixa-tion. O'Donald (1967) computed the progress of sexual selection when females preferred either a dominant or a recessive character dis-favoured by natural selection. Polymorphisms were produced if the selective disadvantage exceeded a certain level. Very strong adverse natural selection was necessary to prevent fixation when females preferred the recessive. Preference for the recessive produces a higher level of linkage disequilibrium and a stronger association between the allele for the preference and the allele for the preferred character. Preference for a recessive is therefore selected more rapidly: a recessive thus gains a greater mating advantage than a dominant. A deleterious recessive character will more often spread through a population by sexual selection than a dominant. It is interesting, therefore, that the red throat of the male stickleback seems to be recessive to non-red. As previously explained (Chapter 1, section 1.4, and Chapter 4, section 4.3) Semler (1971) obtained data on the female preference for sticklebacks with red throats. O'Donald (1973b) estimated that according to the 'C' model with complete mating preferences, 43 per cent of the females exercised a preference for the red males. Only very strong natural selection could overcome the mating advantage that such a preference would give to the red males. Yet some populations do appear to be polymorphic for non-red males (Moodie, 1972). In these popu-lations, sticklebacks are heavily predated by trout. In experimental aquaria trout attacked red sticklebacks much more often then they attacked the non-red sticklebacks. The polymorphism can be ex-plained by the much more intense natural selection against the red males (O'Donald 1973b). The computer models show that in the face of strong adverse natural selection, high levels of female preference will evolve only for a recessive character such as the red throat of the male stickleback.

8.4   Evolution of separate preferences for each genotype

When separate preferences can favour each of the genotypes AA, Aa and aa, the two-locus model is somewhat unrealistic: each of the three alleles, B', B and b, at the second locus now determines mating preferences; and there is no 'null' allele which produces no effect on mating behaviour towards the different male genotypes. However, for greater realism, only a proportion of the females who possess the genotype for a specific preference may actually express their preference. As explained in section 8.3, this proportion represents the penetrance of the genotype that determines the preference. In this particular model of separate preferences the alleles B', B and b have values of penetrance given by α_x, β_x and γ_x. Then, since the dominance relations of the alleles are given by $B' > B > b$, the proportions of females actually expressing preferences are given by

$$\alpha = \alpha_x[\text{frequency of genotypes } B'B', B'B \text{ and } B'b \text{ in females}]$$

$$\beta = \beta_x[\text{frequency of genotypes } BB \text{ and } Bb \text{ in females}]$$

$$\gamma = \gamma_x[\text{frequency of genotype } bb \text{ in females}]$$

Given these preferences, then equation 4.2.2 predicts the following equilibrium frequency of the allele A:

$$p_* = (\alpha + \tfrac{1}{2}\beta)/(\alpha + \beta + \gamma)$$

Of course, as the preferences evolve, the frequencies of B', B and b change; hence α, β, γ and p_* change. As the gene frequencies approach their equilibrium, so the value of p_* converges on the actual frequency p_A. Eventually at equilibrium

$$p_* = p_A$$

and the stable equilibrium state has been attained. Like the model with dominance of the preferred character, the model with separate preferences has the interesting feature that if p_* is close to the actual frequency p_A selection will necessarily be very slow even though the population may be a long way away from the final equilibrium. Since the value of p_* changes as the frequencies of B', B and b change, the actual frequency p_A 'tracks' the frequency p_*. This tracking may produce a very large gene frequency change in time. It may take a very long time, however, if p_A remains close to p_*. In some cases, as I shall show, p_* is always less than p_A and A is eventually eliminated from the population: in other cases p_A may overtake p_*; the population then swings back to a stable polymorphic state. The behaviour of p_A in relation to p_* thus determines the outcome of selection.

In the model with dominance of A, the initial frequencies of both A and B' affected the outcome of selection. In the model with separate preferences for the genotypes AA, Aa and aa, the initial gene frequencies are less important. Table 8.4 gives the final equilibrium frequency of the allele A over a wide range of values in the penetrance of the preferential genotypes. A wide range of initial frequencies gives rise to the same equilibrium. At the same values of penetrance, the equilibria are the same when the initial frequencies of A and B' vary from $p_A=0.1$ to $p_A=0.01$ and $p_B=0.05$ to $p_B=0.005$. Most of the equilibria are still the same if the initial frequencies are raised to $p_A=0.9$ and $p_B=0.85$. The equilibria that sometimes change are those for which α_x, $\beta_x>0.5$ and $\gamma_x<0.3$. At these values of penetrance, the equilibria shown in the table are close to, or at the point of fixation of A: A does in fact reach fixation when $\alpha_x=0.8$, or 0.9. The same equilibria are still reached if the initial frequencies are $p_A=0.9$ and $p_B=0.005$. But if the initial frequencies are $p_A=0.9$ and $p_B=0.05$, then A is lost at the values of penetrance α_x, $\beta_x>0.4$ and $\gamma_x<0.2$. A is also lost at the values α_x, $\beta_x>0.6$ and $\gamma_x=0.2$. All the other equilibria at lower values of α_x and β_x and higher values of γ_x are unchanged given higher initial frequencies of A and B'. The equilibrium frequencies which are altered at the higher initial frequencies are shown in the table in bold type. It seems surprising that the allele A can be eliminated if it starts at a high initial frequency with a greater preference in its favour (determined by the lower values of γ_x and higher frequency of B'), but is not eliminated if it starts at a low frequency with a smaller preference. A hint of the cause of this apparently perverse effect can be obtained by looking at the correlation that develops between the loci, the magnitudes of the preferences that evolve, and the predicted equilibrium value, p_*, calculated from the preferences. Table 8.4(ii) shows what happens during the progress to equilibrium for the case when $\alpha_x=\beta_x=\gamma_x=0.5$. At first, the value of p_* is less than p_A. Consequently p_A declines as it follows p_*: only a few females prefer the AA males. But as a result of linkage disequilibrium, the percentage assortment rises rapidly; because of the association between the alleles A and B' relatively more of the females prefer the AA males; p_A overtakes the value of p_*. Now $p_A<p_*$ and p_A therefore starts to increase. The preference for AA continues to

Table 8.4 *Equilibrium frequencies and the progress to equilibrium when one locus determines the male genotypes and another the females who prefer each genotype separately*

(i) *Equilibrium frequencies of the allele, A, in males*

Penetrance of preference for aa males	Penetrance of preferences for AA and Aa males ($\alpha_x = \beta_x$)								
	0.1	0.2	0.3	0.4	0.5	0.6	0.7	0.8	0.9
0.1	0.560	0.750	0.848	0.888	**0.917**	**0.949**	**0.977**	**0.997**	**1.0**
0.2	0.358	0.567	0.704	0.804	0.883	0.961	**0.997**	**1.0**	**1.0**
0.3	0.241	0.426	0.575	0.698	0.806	0.904	1.0	1.0	1.0
0.4	0.164	0.314	0.454	0.586	0.714	0.843	0.983	1.0	1.0
0.5	0.108	0.221	0.340	0.466	0.603	0.757	0.938	1.0	1.0
0.6	0.065	0.140	0.230	0.337	0.468	0.633	0.851	1.0	1.0
0.7	0.027	0.065	0.118	0.193	0.299	0.455	0.692	1.0	1.0
0.8	0	0	0	0.019	0.073	0.182	0.393	0.836	1.0
0.9	0	0	0	0	0	0	0	0	1.0
1.0	0	0	0	0	0	0	0	0	0

These equilibrium frequencies are stable and reached from a wide range of initial frequencies. They are determined by equation 4.2.2 given the equilibrium frequencies of the females with preferences. The preference for Aa males is always lost: the equilibria are maintained by the preference for AA (determined by B') and the preference for aa (determined by b).

(ii) *Progress to equilibrium when each preferential genotype has a penetrance of 0.5*

Generation	Frequency of allele A p_A	Equilibrium frequency of allele A p_*	Frequency of allele B p_B	Percentage assortment
0	0.0100	0.0070	0.0010	0
1	0.0093	0.0070	0.0010	0.25
2	0.0087	0.0068	0.0010	3.67
5	0.0076	0.0067	0.0010	8.29
10	0.0070	0.0069	0.0012	9.67
20	0.0077	0.0084	0.0018	10.40
50	0.0286	0.0366	0.0167	13.55
100	0.3697	0.3949	0.2352	15.52
200	0.5771	0.5971	0.3744	13.86
500	0.6034	0.6034	0.3967	13.47

In this table, the 'equilibrium frequency of the allele A' is the equilibrium frequency predicted by the proportions of females exercising mating preferences in the particular generation stated. At first, only a very small proportion of the females prefer the AA males and the frequency of A declines. However, the predicted equilibrium frequency does not decline as rapidly as the actual frequency because the alleles A and B' become associated through linkage disequilibrium. After the 10th generation, the predicted equilibrium is overtaken by the actual frequency, which is now less than the predicted frequency and which therefore starts to increase. The allele A eventually reaches a final equilibrium when no further change occurs in the frequencies of the genes for the preferences. The linkage disequilibrium produces a degree of assortment in the preferential matings. The percentage assortment is the frequency, expressed as a percentage of all preferential matings for AA, by which matings of the type $AA \times AA$ exceed an expected frequency calculated on the assumption that preferred and preference genotypes associate at random.

increase maintaining the relation $p_A < p_*$. Therefore p_A continues to track the increasing value of p_*. Finally, when convergence $p_A \to p_*$ has taken place, equilibrium is attained. If the value of p_* were never to be overtaken by p_A, p_A would follow the declining value of p_* until both frequencies had reached zero and A has been eliminated. This is the sequence of events when A starts at a high frequency with a considerable mating preference in its favour.

When $\alpha_x = \beta_x = 0.5$ and $\gamma_x = 0.1$ table 8.4 shows that equilibrium is reached at a frequency $p_A = 0.917$. This equilibrium is always reached if A starts at a low frequency ($p_A = 0.1$, 0.01). It is also reached if A starts at a high frequency and B' starts at a low frequency (e.g. $p_A = 0.9$, $p_B = 0.005$). But A is eliminated if A and B' both start at higher initial frequencies (e.g. $p_A = 0.9$, $p_B = 0.05$). Then the frequency of A never overtakes the value of p_*; p_* continues to decline as B' declines in frequency and A is finally eliminated. Table 8.5 shows the actual and predicted frequencies, p_A and p_*, given the

Table 8.5   *Progress of selection starting at different initial frequencies*

Generation	Initial frequency $p_B = 0.005$			Initial frequency $p_B = 0.05$		
	p_A	p_*	percentage assortment	p_A	p_*	percentage assortment
0	0.9	0.0888	0	0.9	0.4443	0
1	0.8546	0.0888	6.32	0.8519	0.4443	6.41
2	0.8118	0.0886	11.13	0.8089	0.4418	10.89
5	0.6969	0.0874	17.21	0.7035	0.4326	15.61
10	0.5439	0.0837	17.83	0.5853	0.4203	15.65
20	0.3413	0.0754	15.71	0.4698	0.4065	14.72
50	0.1148	0.0581	10.46	0.4008	0.3966	14.02
100	0.0584	0.0566	9.01	0.3955	0.3953	13.96
200	0.1588	0.1993	11.98	0.3940	0.3939	13.96
500	0.6938	0.7007	11.10	0.3875	0.3872	13.93
1000	0.8769	0.8776	5.21	0.3620	0.3610	13.80
2000	0.9092	0.9093	3.38	0.2584	0.2574	12.64
3000	0.9125	0.9125	3.05	0.1872	0.1862	11.52
4000	0.9139	0.9139	2.91	0.1173	0.1163	9.99
5000	0.9147	0.9147	2.83	0.0624	0.0616	8.47
6000	0.9152	0.9152	2.78	0.0294	0.0290	7.42
7000	0.9156	0.9156	2.75	0.0130	0.0127	6.86
8000	0.9158	0.9158	2.73	0.0055	0.0054	6.60
9000	0.9160	0.9160	2.71	0.0023	0.0023	6.49
10000	0.9161	0.9161	2.70	0.0010	0.0009	6.44
20000	0.9166	0.9166	2.66	0	0	6.40

two sets of initial frequencies $p_A = 0.9$, $p_B = 0.005$ and $p_A = 0.9$, $p_B = 0.05$. In the models when either $p_B = 0.005$ or when $p_B = 0.05$ at the start of selection, the value of p_* is low compared to the actual frequency of p_A which therefore declines. In the first few generations the difference between p_A and p_* is considerable. The rate of selection depends on this difference and on the absolute magnitude of the mating preferences. For the first ten generations, the overall rates of selection are similar regardless of whether B' starts at either 0.05 or 0.005. After about twenty generations starting at the frequency $p_B = 0.05$, the declining frequency of A starts to catch up with the declining value of p_* and the rate of selection slows down, becoming very slow after about 100 generations. However throughout the progress of selection the condition $p_A > p_*$ always holds, and A is ultimately eliminated. When B' starts at a frequency of $p_B = 0.005$, the frequency of A continues to decline rapidly until a point is reached at which p_A overtakes the declining value of p_*. This happens because A and B' become slightly more closely correlated, producing a slightly higher percentage assortment. A greater preference evolves for AA relative to the preference for Aa and aa; and hence p_* declines more slowly. A therefore catches up and overtakes p_*. Then $p_A < p_*$. The frequency of A follows p_* to converge at the normal equilibrium frequency. In view of the large number of generations required to produce these changes, too much computation would be needed to establish the range of values over which this difference in the evolutionary process occurs. It does show that when genes at different loci interact, valid predictions can only be made using a specific genetic model: the rate of build up of linkage disequilibrium may often determine the final outcome. Figure 8.2 shows the gene frequency changes of A and B' in the model, given the two different initial frequencies of B', $p_B = 0.05$ and $p_B = 0.005$.

In this general model, when the alleles B', B and b determine separate preferences for the genotypes AA, Aa and aa, the allele B has usually been eliminated by the time equilibrium has been reached: no preference for heterozygotes is left. This is true for almost all values of the penetrance of the preferential genotypes except $\gamma_x = 0.1$. At this value of γ_x, at which 10 per cent of bb females express their preference for aa males, appreciable preference for heterozygotes still remains at equilibrium. At all higher values of γ_x, the preference for heterozygotes is almost or completely eliminated. For example at equilibrium when $\alpha_x = \beta_x = 0.5$ and $\gamma_x = 0.1$, the

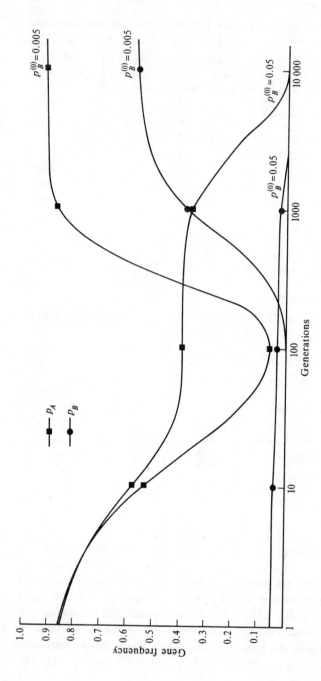

Figure 8.2   Separate preferences for each genotype showing changes in the frequencies of the alleles A and B' when A starts at an initial frequency of $p_A^{(0)} = 0.9$ and B' starts either at $p_B^{(0)} = 0.005$ or $p_B^{(0)} = 0.05$. When B' starts at the lower initial frequency, then the equilibrium shown in table 8.4 is eventually reached after an initial period during which A rapidly declines in frequency. But when B' starts at the higher frequency, then both A and B' are eliminated.

frequencies of females with the preferential genotypes are computed to be

$$(B'B', \ B'B, \ B'b) = 0.81109 \ \text{(preference for } AA)$$

$$(BB, \ Bb) = 0.14146 \ \text{(preference for } Aa)$$

$$(bb) = 0.04745 \ \text{(preference for } aa)$$

Therefore the females who express their preferences are present in the following proportions:

$$\alpha = 0.405545$$

$$\beta = 0.070730$$

$$\gamma = 0.004745$$

giving

$$p_* = (\alpha + \tfrac{1}{2}\beta)/(\alpha + \beta + \gamma)$$

$$= 0.91661$$

as shown in table 8.4(i) of equilibrium frequencies. But at equilibrium, when $\alpha_x = \beta_x = \gamma_x = 0.5$, then

$$(B'B', \ B'B, \ B'b) = 0.60347$$

$$(BB, \ Bb) = 0.00001$$

$$(bb) = 0.39652$$

Therefore

$$\alpha = 0.301735$$

$$\beta = 0.000005$$

$$\gamma = 0.198260$$

and

$$p_* = 0.6035$$

Figures 8.3 and 8.4 show the progress of selection of the alleles A, B', and B in models in which $\alpha_x = \beta_x = \gamma_x$ or $\alpha_x = 2\beta_x = 2\gamma_x$. In figure 8.3 $\alpha_x = \beta_x = \gamma_x = 0.5$: although the final equilibrium frequencies are always the same, with the same correlation between the alleles at the equilibrium, the rate is slightly faster when selection starts at

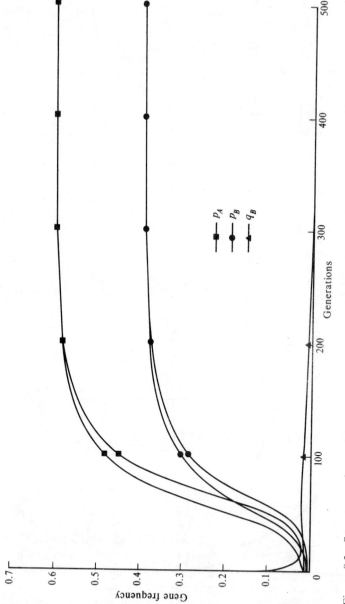

Figure 8.3   Separate preferences for each genotype expressed with 50 per cent penetrance. The model was run with two sets of initial frequencies: (i) $p_A^{(0)} = 0.1$, $p_B^{(0)} = 0.005$, $q_B^{(0)} = 0.01$; (ii) $p_A^{(0)} = 0.01$, $p_B^{(0)} = 0.005$, $q_B^{(0)} = 0.01$. The same equilibria are reached, but at a slightly faster rate from the lower frequency of A since a higher correlation is attained between the loci. The allele B, which has frequency q_B and determines the preference for the heterozygote Aa, is finally eliminated.

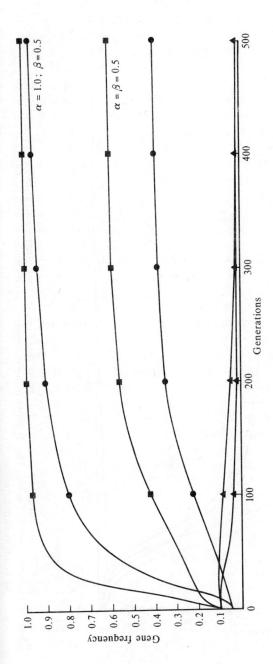

Figure 8.4 Separate preferences for each genotype with different levels of penetrance. When $\alpha_x = 1.0$, $\beta_x = \gamma_x = 0.5$, A is fixed and a eliminated. When $\alpha_x = \beta_x = \gamma_x = 0.5$, then the usual equilibrium frequencies are reached as shown in table 8.4 (ii). The initial frequencies were $p_A^{(0)} = 0.1$ and $p_B^{(0)} = 0.05$. Gene frequencies are shown according to the same key as in figure 8.3.

lower frequencies because the linkage disequilibrium and correlation between alleles builds up more quickly from lower frequencies. In figure 8.4, models in which $\alpha_x = \beta_x = \gamma_x = 0.5$ or $\alpha_x = 2\beta_x = 2\gamma_x = 1.0$ are compared. The rates of selection are of course slower at the lower levels of penetrance of the preferential genotypes, with reduced correlations between loci: when $\alpha_x = 0.5$, polymorphism results; when $\alpha_x = 1.0$, fixation of A is the ultimate outcome.

If $\gamma_x = 0$, then b becomes a 'null' allele with no effect on preference. As expected, A reaches fixation with considerable levels of preference in its favour. As O'Donald (1967) showed in a model in which A was preferred either as a recessive or as a dominant, very rapid selection takes place if females prefer the recessive. If selection starts at low frequencies of both the recessive allele and the allele that determines the mating preference for the recessive, then the preference itself may nearly reach fixation. Figure 8.5 shows what happens if another allele also determines a separate preference for Aa. Two particular cases of the model were used to compute the results shown in the figure. In one case, $\alpha_x = 2\beta_x$, $\gamma_x = 0$. In the other case, $\alpha_x = \beta_x = \gamma_x = 0.5$. When $\alpha_x = 2\beta_x$ and $\gamma_x = 0$, B' determines a preference for AA, B determines a preference for Aa with half the penetrance of B', and b is a null allele. The allele A always then reaches fixation in spite of the fact that B' starts at a much lower frequency than B giving a greater initial preference for Aa than for AA. Although B increases in frequency during the first fifty generations, it becomes only slightly correlated with A: B' becomes much more closely correlated. Eventually the correlation of B with A becomes negative and ultimately B is eliminated. Thus no final preference remains for Aa and $\beta = 0$. Since from the start in this model we put $\gamma = 0$, then

$$p_* = (\alpha + \tfrac{1}{2}\beta)/(\alpha + \beta)$$

$$= \alpha/\alpha$$

$$= 1$$

and A becomes fixed as shown in the figure. When $\alpha_x = \beta_x = \gamma_x = 0.5$, then as before we have

$$p_* = 0.603$$

The rates of selection towards this equilibrium are shown in the figure for comparison with the case when $\gamma_x = 0$.

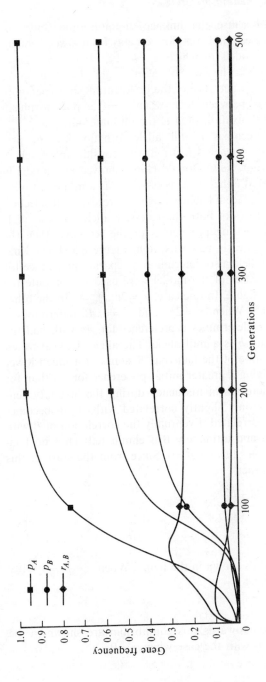

Figure 8.5   Separate preferences for genotypes AA and Aa ($\alpha_x = 2\beta_x = 0.5$, $\gamma_x = 0$) compared with separate preferences for each of the genotypes AA, Aa and aa ($\alpha_x = \beta_x = \gamma_x = 0.5$). In both runs of the model the initial frequencies were $p_A^{(0)} = 0.001$, $p_B^{(0)} = 0.001$ and $q_B^{(0)} = 0.005$. The preference for the heterozygote determined by the allele B at frequency q_B, was therefore initially greater than the preference for the homozygote AA determined by the allele B' at frequency p_B. When $\alpha_x = 2\beta_x = 0.5$, $\gamma_x = 0$, A reaches fixation, B is maintained at $p_B = 0.062$, and the correlation between the loci is reduced ultimately to zero. When $\alpha_x = \beta_x = \gamma_x$, then the equilibria reached are those given in table 8.4(ii) with $p_A = 0.603$, $p_B = 0.397$ and $r_{A,B} = 0.241$. The preference for Aa is always eliminated.

8.5 Evolution of partial preferences for dominant or recessive characters

In the models of sections 8.3 and 8.4, the penetrance of the genotypes that determined the female mating preferences was assumed to be constant: a particular frequency of an allele for a mating preference gave rise to a constant proportion of preferential matings; the expression of the mating preference did not depend on the frequency of the preferred males. The penetrance might also be a function of the frequency of the preferred males, however, as in the 'P' and 'PF' models. In the 'P' models, a female mates at random if she does not encounter a male she prefers on n successive occasions.

If a proportion α of the females prefer A males and γ prefer B males, then the probabilities of preferential matings with A and B males are given by $\alpha(1-y^n)$ and $\gamma(1-x^n)$ where x and y are the frequencies of the phenotypes A and B. Thus we can substitute these expressions for α and γ in equations 3.1.1 for the frequencies of matings with A and B phenotypes of males. Equations 3.1.1 which are

$$P_T(A) = \alpha + x(1 - \alpha - \gamma)$$

$$P_T(B) = \gamma + y(1 - \alpha - \gamma)$$

thus become the equations 3.2.1 as follows:

$$P_T(A) = \alpha(1 - y^n) + x(1 - \alpha + \alpha y^n - \gamma + \gamma x^n)$$

$$= \alpha(1 - y^{n+1}) + \gamma x^{n+1} + x(1 - \alpha - \gamma)$$

$$P_T(B) = \alpha y^{n+1} + \gamma(1 - x^{n+1}) + y(1 - \alpha - \gamma)$$

To transform the computer models of 8.4 into the 'P' models with partial expression of preference, it is merely necessary to substitute for the constant penetrance of the 'C' models, a level of penetrance that varies in proportion to $1 - (1 - x)^n$ where x is the frequency of the phenotype or genotype that is the object of the preference.

We should expect that the results of selection for mating preferences in the 'P' model will converge on those in the 'C' model as n is increased. Even so, if the preferred character is very rare, only a very small proportion of females will mate according to their preference. At the start of sexual selection for a new allele that improves the development of the preferred character, the rate of selection should therefore be much slower in the 'P' model than in the 'C' model. Linkage disequilibrium will be slower to build up and the mating preferences should evolve more slowly in the 'P' model. We should also expect that mating preferences would be less likely to evolve to maintain a polymorphism: the loss of frequency-dependent mating success at the lower values of n will tend to produce a consistent overall advantage for one particular phenotype.

I have investigated the outcome of the evolution of mating preferences in the 'P' models selecting either for dominant and recessive phenotypes or for each genotype separately. Table 8.2 in section 8.3 gave equilibrium frequencies of the allele A for particular values of α_x and γ_x and different initial frequencies of A, B' and B. As in the models of section 8.3, so also in the models of this section A is dominant to a; B', which is dominant to B, determines the preference for A; and B determines the preference for a. Table 8.6 shows the equilibrium frequencies of A and the residual percentage assortment of B' among A phenotypes at the point when equilibrium has been reached. Unlike equilibria in the model of separate preferences for each genotype, these equilibria are always determined by the initial gene frequency. This was shown in the table of equilibria in the 'C' model for sexual selection of a dominant and recessive phenotype (table 8.2, section 8.3). All the equilibria for the 'P' models (table 8.6) were reached from the same set of initial frequencies of A, B' and B: $p_A = 0.1$, $p_B = 0.05$ and $q_B = 0.1$. The table shows that the 'P' model behaves according to our expectations. At a high value of n ($n = 10$), the equilibria are similar to those of the 'C' model. But as n is reduced the equilibrium frequencies tend towards fixation of A at the lower values of γ_x and towards fixation of a at the higher values of γ_x. When $n = 2$, polymorphisms are maintained only within the range of values $0.1 < \gamma_x < 0.5$. When $n = 1$, all the frequency-dependent advantage of being rare has been lost. Then A is fixed if $\gamma_x \leq 0.2$ and a is fixed if $\gamma_x > 0.2$. Since in all these results $\alpha_x = 0.5$, this shows that the recessive a still has an advantage over A even when the penetrance in its mating preference is much less than the penetrance in the mating preference of the dominant. This may have been expected from previous results (O'Donald, 1967) showing that higher levels of preference can evolve in favour of a recessive.

The rate of selection of mating preferences depends on the linkage disequilibrium between the loci, which O'Donald (1967) found to be generally higher when the preferred phenotype was

Table 8.6 Equilibrium frequencies in the 'P' model when A is a dominant and a is a recessive

Value of penetrance of aa	Equilibrium frequency and percentage assortment									
	'C' model		'P' model ($n=10$)		'P' model ($n=5$)		'P' model ($n=2$)		'P' Model ($n=1$)	
γ_x	p_A	% assort	p_A	% assort	p_A	% assort	p_A	% assort	p_A	% assort
0.1	0.5507	2.96	0.5784	2.49	0.7355	0.70	1.0	0	1.0	0
0.2	0.4143	5.99	0.4167	5.92	0.4450	5.18	0.9202	0.02	1.0	0
0.3	0.3302	8.51	0.3297	8.52	0.3320	8.29	0.3638	6.23	0	0
0.4	0.2717	10.67	0.2706	10.59	0.2597	10.48	0.1209	6.95	0	0
0.5	0.2286	12.27	0.2268	12.27	0.2049	11.91	0	0	0	0
0.6	0.1956	13.68	0.1925	13.62	0.1592	12.59	0	0	0	0
0.7	0.1697	14.86	0.1644	14.67	0.1188	12.36	0	0	0	0
0.8	0.1489	15.86	0.1408	15.43	0.0811	10.94	0	0	0	0
0.9	0.1318	16.71	0.1202	15.89	0.0446	7.75	0	0	0	0
1.0	0.1177	17.43	0.1019	16.02	0.0079	1.78	0	0	0	0

For each set of values the initial frequencies were set at $p_A = 0.1$, $p_B = 0.05$ and $q_B = 0.1$ where p_A, p_B and q_B are the frequencies of the alleles A, B and B; the penetrance of A was always $\alpha_x = 0.5$.

recessive. This produces the faster selection of the females who prefer the recessive in the 'C' models. How will selection for a recessive compare with that for a dominant in the 'P' models? In answering this question, we may expect to find less difference in the rates of selection of dominants and recessives in the 'P' models: if a dominant and a recessive both have similar gene frequencies, the recessive phenotype will be much rarer than the dominant phenotype and a much smaller proportion of females will therefore express their preference for the recessive. In this respect, the evolution of mating preferences may proceed at very different rates and to very different levels in the 'P' and 'C' models.

Evolution of preference for a dominant character

Figure 8.6 shows the frequencies of the allele A and figure 8.7 shows the corresponding frequencies of the allele B' in the 'C' and 'P' models given the same initial frequencies $p_A = p_B = 0.01$. In the 'C' model selection is much faster and B' rises to a higher final value than in the 'P' models with various values of n. In the 'C' model B' shows nearly a five-fold increase in frequency: in the 'P' model with $n = 1$ it only increases about 1.5 times. This is also reflected in the greater correlation between loci that develops in the 'C' model and determines the more rapid selection of the mating preference.

As I had previously noted (O'Donald, 1967), there is greater selection for the mating preference when initial frequencies are lower. Table 8.7 shows the final frequencies attained by B' from various initial frequencies. The greatest proportional increase in p_B occurs when A and B' both start at the lower frequencies: the smallest increase occurs when they both start at the higher frequencies. Of course, selection is faster when B' starts at the higher frequency. If A and B' both start at frequencies $p_A = p_B = 0.0001$, then selection hardly gets started in the first 2000 generations and always stays very slow indeed: B' only increases slightly. If A and B' were both to start at about the mutation rate of $\mu = 0.00001$, then selection would hardly get started in the first 20 000 generations. The selection is negligible at these low initial frequencies because females virtually never encounter any male they prefer: a negligible proportion express their preference. As Fisher suggested, preferences start to evolve only if females prefer those male phenotypes that are advantageous in natural selection. Females who express

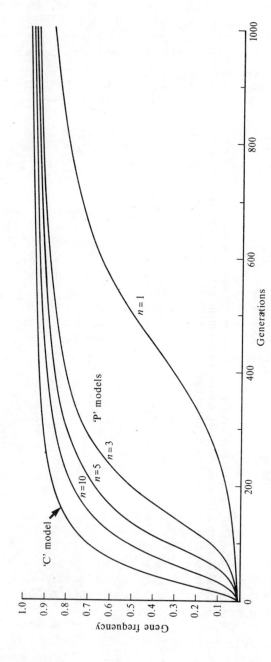

Figure 8.6 Complete or partial expression of preference for a dominant allele. The initial frequencies were $p_A^{(0)} = p_B^{(0)} = 0.01$. The preference for the dominant A phenotype, determined by the allele B', was fully penetrant with $\alpha_x = 1.0$. The figure shows the frequencies of the allele A during the progress of selection towards fixation of A. As expected, when females with preferences mate at random after fewer encounters, the rate of selection is reduced.

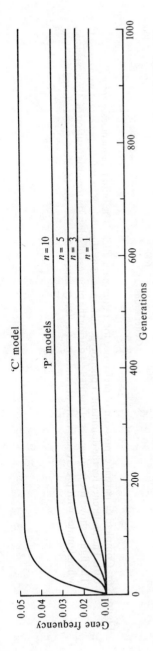

Figure 8.7 Complete or partial expression of preference for a dominant allele. The figure shows the frequencies of the allel B', which determines the preferences for the dominant A phenotype, during the progress of A towards fixation. Other details are given in the legend to figure 8.6.

Table 8.7 *Final frequencies of the mating preference gene, B', after selection of a dominant character to near fixation*

Model	Initial gene frequencies in the model							
	$p_A=0.1$ $p_B=0.1$		$p_A=0.1$ $p_B=0.01$		$p_A=0.01$ $p_B=0.1$		$p_A=0.01$ $p_B=0.01$	
	p_B	T	p_B	T	p_B	T	p_B	T
'C' Model	0.1673	700	0.0205	5000	0.2364	500	0.0488	2000
'P' Model $n=10$	0.1670	700	0.0203	5000	0.2364	500	0.0336	3000
'P' Model $n=5$	0.1637	700	0.0193	5000	0.2106	500	0.0267	4000
'P' Model $n=3$	0.1572	700	0.0180	6000	0.1875	600	0.0224	5000
'P' Model $n=1$ ·	0.1351	800	0.0146	7000	0.1445	800	0.0158	7000

In all the models analysed in this table, the values of p_B shown had reached their final equilibrium by the time p_A had reached 0.99. By this time selection for the dominant A had become very slow. The value given for T is the approximate number of generations required to reach $p_A=0.99$ thus giving an indication of the rate of selection. T is given to the nearest 100 or 1000 generations.

such preferences gain their initial advantage only through their sons who possess the advantageous character. Once the mating preference has increased in frequency itself, then it adds to the selective advantage of the males in promoting Fisher's 'runaway process'. In section 8.7, I analyse models incorporating natural selection. These models demonstrate the possible sequence of events in the whole of the evolutionary process of sexual selection, starting with selection in favour of females who may be able to distinguish and mate with the advantageous male phenotypes and concluding when further sexual selection is halted by adverse natural selection. When the preference has increased in frequency, it will fairly rapidly select for rare alleles that enhance the development of the characters that the females use to discriminate between the males. If there is a 'peak shift' (see section 8.1) in the female responses in favour of more exaggerated developments of the character, this may perhaps also increase the penetrance of the mating preferences as they evolve. The table shows that once the allele B' has spread to about 20 per cent of the females (i.e. a preference determined by a dominant allele at a frequency of $p_B=0.1$) further alleles that enhance the preferred character may be selected rapidly.

Evolution of preference for a recessive character

Since a recessive starts at a lower phenotypic frequency than a dominant given the same initial gene frequency for both, selection for a recessive must necessarily start more slowly in the 'P' models but should finish more quickly than selection for a dominant. In the computations, ultimate fixation can virtually be reached. Figure 8.8 shows the frequencies of the allele a and figure 8.9 shows the corresponding frequencies of the allele B in the 'C' and 'P' models given the initial frequencies $q_A = q_B = 0.01$. Selection now takes a very long time in the 'P' models, but is very quick in the 'C' model. This points to the sharp contrast we had expected between selection for dominants on the one hand and recessives on the other. In the 'C' models, selection of a recessive is much faster than selection of a dominant, with high levels of linkage disequilibrium being produced. But in the 'P' models, selection for dominants is similar to selection for recessives, though generally somewhat higher levels of preference always evolve if a recessive is selected. Table 8.8 shows the great contrast between the 'C' and 'P' models in the evolution of the preference for a recessive. At the same time, the gene frequencies of B in this table should be compared with the gene frequencies of B' for the preference for the dominant given in the previous table 8.7. At the lower frequency of a, the selection takes place very slowly in the 'P' models at first; the preference gene for the recessive is selected to an only slightly higher frequency than the preference gene for the dominant. The slow initial selection in the 'P' models is caused by the improbability that a female should meet one of the recessive males she prefers. In these models at the higher frequencies of a, however, selection of the recessive is considerably faster than selection of the dominant and the preference evolves to higher levels: the females have a higher chance of meeting the males they prefer, and thus of expressing their mating preference. Figure 8.10 shows the effect of the model and the initial frequency on the rate of selection for a dominant; figure 8.11 shows the corresponding effects on the rate of selection for a recessive. In the 'C' model, when a recessive is selected with initial frequencies $q_A = q_B = 0.1$, the correlation between a and B has a maximum value $\rho_{aB} = 0.243$ in generation four. In the same model, when the recessive is selected with initial frequencies $q_A = q_B = 0.01$, the correlation has a maximum $\rho_{aB} = 0.405$ in generation seven. This explains the rapid selection of the preference for the recessive in the 'C' model, particularly when initial frequencies are low and a higher

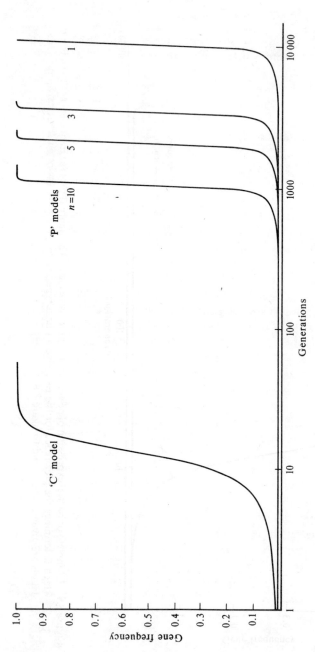

Figure 8.8 Complete or partial expression of preference for a recessive allele. The initial frequencies of the alleles for the recessive a and the preference B were $q_A^{(0)} = q_B^{(0)} = 0.01$. The preference for the recessive a phenotype was fully penetrant with $\gamma_x = 1.0$. The figure shows the frequencies of the allele a during the progress of selection towards fixation of a. Selection is very much more rapid in the 'C' model than in the 'P' models; in the 'P' model in which females with preferences mate at random after just one encounter, selection is initially very slow indeed. The generations have been plotted on a logarithmic scale in order to compare the models over the great difference in time at which a increases in frequency to fixation.

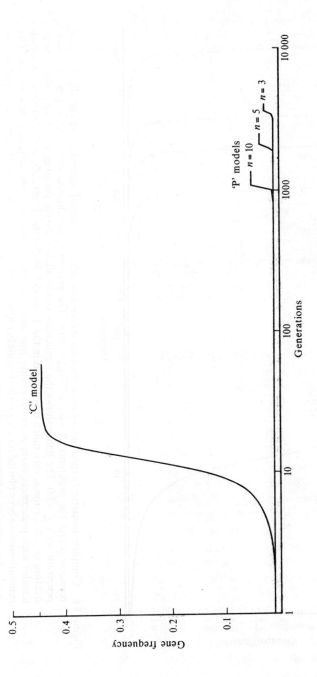

Figure 8.9 Complete or partial expression of preference for a recessive allele. The figure shows the frequencies of the allele B, which determines the preference for the recessive a phenotypes, during the progress of a towards fixation. Other details are given in the legend to figure 8.8.

Table 8.8   *Final frequencies of the mating preference gene, B, after selection of a recessive to complete fixation*

Model	Initial gene frequencies in the model							
	$q_A=0.1$ $q_B=0.1$		$q_A=0.1$ $q_B=0.01$		$q_A=0.01$ $q_B=0.1$		$q_A=0.01$ $q_B=0.01$	
	q_B	T	q_B	T	q_B	T	q_B	T
'C' model	0.3348	60	0.1689	121	0.4196	51	0.4463	56
'P' model $n=10$	0.2826	77	0.0497	423	0.2916	175	0.0522	1343
'P' model $n=5$	0.2322	100	0.0333	669	0.2359	297	0.0342	2531
'P' model $n=3$	0.1991	127	0.0257	938	0.2011	455	0.0261	4052
'P' model $n=1$	0.1470	237	0.0164	2016	0.1476	1221	0.0165	11385

In all the models analysed in this table, the values of q_B are those given at the final equilibrium when the frequency of a had reached $q_A=1.0$ to a level of numerical accuracy given by double-precision arithmetic in the IBM 370/165 Computer.

association develops between the loci. But in the 'P' model with $n=3$, then starting at the higher frequencies $q_A=q_B=0.1$ the maximum value $\rho_{aB}=0.090$ is reached in generation forty; and starting at the lower frequencies $q_A=q_B=0.01$ the maximum value $\rho_{aB}=0.003$ is reached only by about generation 3000. These considerable differences in the correlations between the loci explain the great differences between the 'C' and 'P' models in the rates of selection for a recessive character.

If the preferred character has already made some progress in its evolution (presumably it would do so by natural selection on the premises of Fisher's theory) then the preference would have little effect either on the selection for a recessive or on the selection for a dominant. Tables 8.7 and 8.8 show that given the initial frequencies $p_A=0.1$, $p_B=0.01$ (dominant character preferred) or $q_A=0.1$, $q_B=0.01$ (recessive character preferred) the sexual selection is always slow in the 'P' models and would probably be relatively unimportant compared to the natural selection that might favour the preferred character. On the other hand, if the preferred character and preference both evolve together as Fisher suggested, at first by natural selection and later to a greater extent by sexual selection as

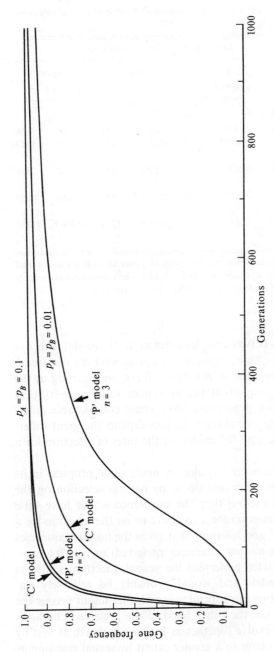

Figure 8.10 Complete or partial expression of preference for a dominant allele with different initial frequencies. The initial frequencies determine the preference for the dominant A phenotype with $\alpha_x = 1.0$. At the higher initial frequencies, selection is almost as rapid in the 'P' model as in the 'C' model. At the lower initial frequencies, selection is slower in the 'C' model and slower still in the 'P' model compared to selection at the higher initial frequencies.

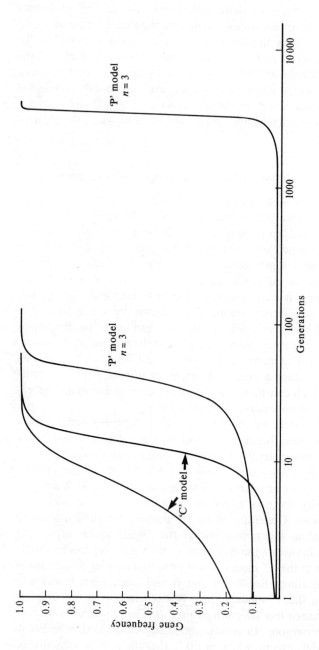

Figure 8.11 Complete or partial expression of preference for a recessive allele with different initial frequencies. The model was run with two sets of initial frequencies: (i) $q_A^{(0)} = q_B^{(0)} = 0.1$; (ii) $q_A^{(0)} = q_B^{(0)} = 0.01$. The allele B determines the preference for the recessive a phenotype with $\gamma_x = 1.0$. The 'C' model produces rapid selection at both the higher and lower initial frequencies. The 'P' model produces very slow selection at the lower initial frequencies.

the preference becomes more widespread, the mating preference would reach an appreciable frequency by the time fixation of A or a occurred. Provided the females with preferences continue to be more responsive to males with more elaborate developments of the preferred character, an allele either at the A locus, or at another locus, which enhances the preferred character, would obviously be selected rapidly by sexual selection; the preference itself would also increase, adding to the sexual selective advantage and outweighing the effects of natural selection.

8.6 Evolution of partial preferences for each genotype

We have already seen (section 8.4) that when the alleles B', B and b determine separate preferences for each of the genotypes AA, Aa and aa, then polymorphisms are produced over a wide range of values in the penetrance with which the preferences are expressed by the females. Usually, the same equilibria are reached in spite of wide variation in the initial frequencies. However, if A starts at a high frequency, it may sometimes be lost completely. When this occurs, the equilibrium frequency predicted by the mating preferences that have evolved is always less than the actual frequency. The actual frequency therefore follows the predicted frequency down to zero. The mating preference for AA never evolves quickly enough to produce a predicted frequency greater than the actual frequency which can then start to increase again eventually reaching the general polymorphic equilibrium.

In the 'P' models, we should expect that stable equilibria will be reached less often because the loci will become less closely associated at lower levels of linkage disequilibrium and with a reduced level of assortative mating. Table 8.9 shows how clearly these expectations are realized. The equilibria of the 'C' model are compared with the final effects of selection in the 'P' models with $n = 10$ and $n = 1$. Generally, in the 'P' models, the allele a tends to become fixed in the population at the higher values of γ_x. The correlations between A and B' and a and b are also less in the 'P' models than in the 'C' models, giving rise to a smaller percentage of assortative matings. Because A and B' and a and b are less closely correlated in the 'P' models, there is less selection in favour of B' and b and hence less selection against B. Therefore the preference for the heterozygote Aa is not completely eliminated; whereas in the 'C' model, except when $\gamma_x = 0.1$, the allele B is virtually or

Table 8.9 *Equilibrium frequencies after the evolution of separate preferences for each of the genotypes* AA, Aa *and* aa *(the penetrance in the mating preferences for* AA *and* Aa *was* $\alpha_x = \beta_x = 0.5$ *throughout these computations)*

Penetrance in mating preference for aa γ_x	Model used in computation					
	'C' Model		'P' Model $n = 10$		'P' Model $n = 1$	
	p_A	percentage assortment	p_A	percentage assortment	p_A	percentage assortment
0.1	0.917	2.74	0.847	5.78	0.378	1.67
0.2	0.883	4.66	0.891	3.97	0.104	0.04
0.3	0.806	7.97	0.895	4.11	0.239	0.47
0.4	0.714	10.81	0.168	2.13	0	0
0.5	0.603	13.47	0.099	0.50	0	0
0.6	0.468	15.77	0.060	0.09	0	0
0.7	0.299	17.43	0.041	0.01	0	0
0.8	0.073	18.18	0.040	0.01	0	0
0.9	0	1908	0.003	0	0	0
1.0	0	20.20	0	0	0	0

Selection is very slow in the 'P' models when $\gamma_x \leq 0.3$. The values shown in the table may still have a long way to go before reaching the final equilibrium: they represent not the true equilibria but the frequencies reached after 20 000 generations; at this point computation was stopped.

completely eliminated, leaving hardly any residual preference for *Aa* or none at all. The preference for heterozygotes is generally reduced, however, and particularly in the 'P' models with small values of *n* this gives rise to unstable equilibria. In the absence of any preference for *Aa*, the polymorphic equilibria are stable only in the 'C' model and in the 'P' model with high values of *n*. This is shown in table 5.5 (ii) Chapter 5, section 5.2, where the equilibria are unstable unless $n > 10$–15. At these high values of *n* the 'P' and 'C' models become very similar in their results.

8.7   Natural selection and the evolution of mating preferences

Natural selection must now be incorporated in the model of the evolution of mating preferences in order to analyse the sequence of events in Fisher's general theory of sexual selection (section 8.1). We have already noted (section 8.5) that natural selection will be an essential factor in the model if mating preferences are ever to evolve in a reasonable period of biological time. Once they have evolved, however, the selective enhancement of sexual display may proceed largely by sexual selection with natural selection having relatively little effect. Only when the characters for sexual display have evolved to extremes will adverse natural selection have become powerful enough to stop the runaway process of sexual selection. The 'P' model, with natural selection of the genotypes *AA*, *Aa* and *aa*, can be used to exemplify and analyse these three stages in the overall evolutionary process.

In the first stage of the process we assume that a character, which the females can recognize and distinguish, is pleiotropically associated with higher fitness in the males. At first, both the males with the character, and the females who can recognize and respond to them, are rare in the population. The alleles, at the loci which determine the character and mating preferences, may occur at frequencies not much greater than the mutation rate. Let us now see what happens in the course of selection of alleles *A* and *B'* (which determine a dominant character and the preference in its favour) or alleles *a* and *B* (which determine a recessive character and the preference in its favour). The rate at which mating preferences can evolve will determine the importance of sexual selection in the evolution of sexual dimorphism. Starting at low initial frequencies, we can calculate how far the mating preference alleles will be selected while an allele that determines the preferred character spreads through the population. Table 8.10 gives the results of these calculations for models in which the preferred character is dominant; table 8.11 gives the results for the recessive. The dominant character is always selected very slowly as it nears fixation. No time to actual fixation can be given. The frequencies of *B'* and numbers of generations, *T*, are therefore shown at the time when p_A reaches 0.95. By this time no further change in p_B takes place. The mating preference gene reaches about the same frequency whether or not natural selection promotes the spread of *A*. But with the aid of natural selection the time taken is more than twenty times faster in the 'P' model with $n = 10$ and forty times faster in the 'P' model with $n = 3$. Without natural selection, the evolution hardly gets started before 1000 generations have passed.

The difference in rate is even more striking when a recessive character is selected. Without the aid of natural selection, no perceptible changes occur for thousands of generations. This is a consequence of the extreme rarity of the recessive phenotypes:

Table 8.10   *Initial selection of the mating preference for a dominant character that spreads through the population (initial frequencies $p_A = p_B = 0.001$)*

| | Model used in computation | | | |
| | 'P' model $n = 10$ | | 'P' model $n = 3$ | |
	p_B	T	p_B	T
Natural selection absent	0.0038	7000	0.0023	12 000
Natural selection in favour of A	0.0039	300	0.0023	300

In these computations, females carrying the mating preference gene B' prefer males with the dominant character A. The initial frequencies, p_A and p_B, of the alleles A and B', are given in the heading to the table. Shown in the table are the final frequencies of B' after T generations during which A had reached a frequency $p_A = 0.95$. By this time no further change in the frequency of B' takes place. If natural selection also favoured A, it was determined by a selective disadvantage against aa of $t = 0.1$.

females have virtually no opportunity to express any preference for the recessive at initial gene frequencies $q_A = q_B = 0.001$ of the alleles a and B. If natural selection favours the aa genotype, however, the allele a ultimately reaches fixation (since it is selected much more rapidly than a dominant as it nears fixation), and B reaches a higher final frequency than B'. Approximately, B' undergoes a two-fold to four-fold increase in frequency and B a three-fold to six-fold increase. Therefore, if a small proportion of females can recognize and respond more rapidly to males with an advantageous character, females with this mating preference will increase their frequency several times while the advantageous character spreads through the population.

During the next stage in this evolutionary process, the character is improved and enhanced for sexual display: the females' responsiveness increases with the development of the character. New alleles, starting from low frequencies and enhancing the character, are now selected. Tables 8.12 and 8.13 show how the frequencies of the mating preference alleles B' or B change during the selection of the dominant A or the recessive a. The alleles A and a are now new alleles, perhaps at new loci, that enhance the development of the original character. The alleles B' and B start at higher frequen-

Table 8.11 *Initial selection of the mating preference for a recessive character that spreads through the population*

(i) *Initial frequencies* $q_A = q_B = 0.001$

	Model used in computation			
	'P' model $n=10$		'P' model $n=3$	
	q_B	T	q_B	T
Natural selection absent	0.001	–	0.001	–
Natural selection in favour of *aa*	0.0061	8390	0.0027	8950

(ii) *Initial frequencies* $q_A = q_B = 0.01$

	Model used in computation			
	'P' model $n=10$		'P' model $n=3$	
	q_B	T	q_B	T
Natural selection absent	0.0522	1343	0.0261	4052
Natural selection in favour of *aa*	0.0555	576	0.0266	848

In these computations, females carrying the mating preference gene B prefer males with the recessive genotype *aa*. The initial frequencies, q_A and q_B, of the alleles *a* and *B* are given in the headings to the tables. The final frequency of B is then shown after T generations at the point of fixation of *a*. If natural selection also favoured *aa*, it was determined by a selective disadvantage against A of $s=0.1$.

cies, since the mating preferences have already made considerable evolutionary progress. The further selection of the dominant is still considerably faster than the recessive, but natural selection makes much less of a difference once the mating preferences have already undergone initial selection. The mating preferences continue to progress, increasing about two or three times in frequency. But the rate of selection, which was so greatly increased by natural selection when the mating preference genes were low in frequency, is now only slightly raised by natural selection. When low in frequency, the mating preference added little or nothing to the selection of the preferred character though in association with the preferred character, the genes for the preference increased substantially in frequency. Having increased, however, the mating preference will mainly be responsible for the further evolution of

Evolution of mating preferences

Table 8.12 *Further selection of the mating preference for a dominant character that spreads through the population (initial frequencies $P_A=0.001$, $p_B=0.1$)*

	Model used in computation			
	'P' model $n=10$		'P' model $n=3$	
	p_B	T	p_B	T
Natural selection absent	0.2589	500	0.1920	600
Natural selection in favour of A	0.2587	400	0.1914	400

The initial frequencies, p_A and p_B, of the alleles A and B' are given in the heading to the table. Shown in the table are the final frequencies of B' after T generations during which A had reached a frequency $p_A=0.99$. This frequency is higher than the corresponding frequency $p_A=0.95$ used to measure the number of generations during the initial selection of the mating preference (table 8.10). The higher final frequency $p_A=0.99$ was chosen because at $p_A=0.95$, B' was still evolving in the model of the further selection of the mating preference.

Table 8.13 *Further selection of the mating preference for a recessive character that spreads through the population (initial frequencies $q_A=0.001$, $q_B=0.1$)*

	Model used in computation			
	'P' model $n=10$		'P' model $n=3$	
	q_B	T	q_B	T
Natural selection absent	0.2917	1137	0.2011	3657
Natural selection in favour of aa	0.2986	922	0.2031	2433

The initial frequencies, q_A and q_B, of the alleles a and B are given in the heading to the table. The final frequency of B is then shown after T generations at the point of fixation of a.

the character: sexual selection will enhance the development of the character with only slight assistance from natural selection. Figure 8.12 illustrates the effect of natural selection on the rates of selection of a dominant character before and after selection of the mating preference has taken place.

In the final stage of evolution, the runaway process of sexual

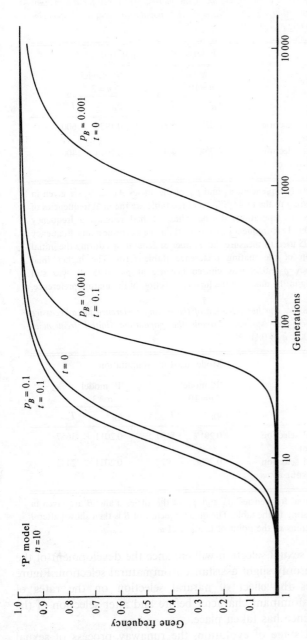

Figure 8.12 Partial preference for a dominant allele with natural selection and different initial frequencies. The preference for the dominant A phenotype, determined by the allele B', was fully penetrant with $\alpha_x = 1.0$. The model was run with two sets of initial frequencies: (i) $p_A^{(0)} = p_B^{(0)} = 0.001$; (ii) $p_A^{(0)} = 0.001$, $p_B^{(0)} = 0.1$. The figure shows that natural selection, which has little effect on the rapid selection at the higher initial frequency of B', has the dominant effect on selection at the lower initial frequency of B' when it produces roughly a ten-fold increase in the rate of selection.

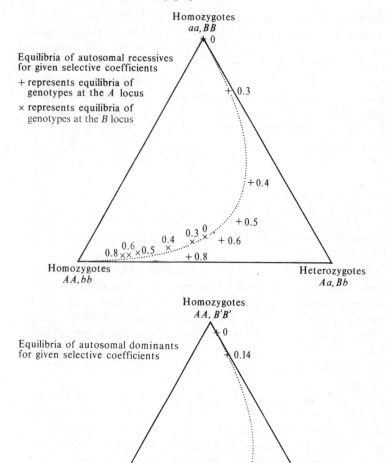

Figure 8.13 Preferences for dominants or recessives opposed by natural selection producing stable polymorphism. The allele *B* determines a complete preference for a recessive allele *a*; *B'* determines a complete preference for a dominant allele *A*. The coefficients of natural selection acting against the preferred recessives or dominants are shown in the figure.

selection is brought to a halt. The character has developed to such an extreme that it has now become deleterious. Natural selection has turned against the character so much that the sexual selection is outweighed. In the 'C' models, very strong natural selection is necessary to stop the further sexual selection of recessive alleles: much less strong natural selection stops the sexual selection of dominant alleles (O'Donald, 1967). A wide range of selective values gives rise to stable polymorphisms. This is a consequence of the strong frequency-dependence of the sexual advantage in the 'C' models: while the preferred males are rare, their very great advantage in sexual selection will outweigh very severe natural selection against them; but since their sexual advantage declines rapidly with increasing frequency, a balance will eventually be reached between the forces of sexual and natural selection. Figure 8.13 (taken from O'Donald, 1967) shows the equilibria at various intensities of natural selection.

In the 'P' models, in which the sexual advantage declines more slowly with frequency, we should expect that polymorphisms would become established less often than in the 'C' model. In the 'P' model with $n = 1$, this frequency-dependence is lost completely, and polymorphisms can never be established. The character will continue to develop so long as the sexual selection in its favour is stronger than the natural selection against it. Past this point of development, no more new alleles will be sexually selected. For larger values of n some polymorphisms may become established; but in contrast with the 'C' models they will not be established at a balance with very intense natural selection, for the initial sexual advantage will not be so great as to outweigh the natural selection. Polymorphisms can only become established over a much more limited range of selective values.

9

Assortative mating and sexual selection

9.1 Preferences for dominant and recessive characters

In the introductory chapter of this book (section 1.5), I drew a distinction between sexual selection and assortative mating. In an organism that mates monogamously, with no assortment of preference within preferred genotypes or phenotypes, the overall frequencies become the same as those of random mating (table 3.1). Deviations from random mating occur only with assortment. Non-assortative or random preferential mating is thus clearly distinct from assortative preferential mating. In monogamous organisms, preferential mating gives rise to sexual selection as defined by Darwin only if some matings are less fertile than others or if some individuals are left unable to find mates. Darwin's own theory analysed in Chapter 7, assumes that since the preferred males are more readily chosen as mates, they then produce more offspring because earlier matings are more successful than later ones. As we have shown, Darwin's assumptions about breeding ecology are generally true for non-passerine birds.

In polygynous organisms, strong sexual selection is the rule with both types of preferential mating. But mating preferences, whether they be randomly distributed or assortative, do not by themselves produce selection: the selection acting on preferred individuals is a consequence of the breeding ecology of the organism and whether the mating system is polygynous or monogamous. The nature of the mating preferences should therefore be distinguished from the mechanisms that give rise to selection.

Karlin and O'Donald (1978) analysed models of polygynous organisms with parameters for both random and assortative preferential mating. They used the term sexual selection to describe mate selection by the randomly distributed preferences and thus distinguished this process from that of assortative mating.

Table 9.1 *Mating frequencies in the combined model with preferences for both dominant and recessive characters*

Mating types	Components of matings		
	Random preferential mating	Assortative preferential mating	Random mating
$AA \times AA$	$\alpha u^2/(1-w)$	$au^2/(1-w)$	$u^2(1-\alpha-\gamma-a)$
$AA \times Aa$	$2\alpha uv/(1-w)$	$2auv/(1-w)$	$2uv(1-\alpha-\gamma-a)$
$AA \times aa$	$\alpha uw/(1-w)+\gamma u$	0	$uw(2-2\alpha-2\gamma-a-c)$
$Aa \times Aa$	$\alpha v^2/(1-w)$	$av^2/(1-w)$	$v^2(1-\alpha-\gamma-a)$
$Aa \times aa$	$\alpha vw/(1-w)+\gamma v$	0	$vw(2-2\alpha-2\gamma-a-c)$
$aa \times aa$	γw	cw	$w^2(1-\alpha-\gamma-c)$

In this table α and γ are the parameters for random preferential mating and a and c the parameters for assortative mating of the characters A (genotypes AA and Aa) and a (genotype aa).

Although this terminology was useful in discussing the selective processes that occur in the combined model, it is a restricted use of the term 'sexual selection' compared to Darwin's usage: the preferred males are sexually selected as a result of both the random and the assortative preferences in Karlin and O'Donald's models. Hence in section 1.5 I put forward the terms assortative preferential mating and random preferential mating to describe matings determined by preferences acting with and without assortment. Use of these terms thus separates the behavioural mechanisms that determine the frequencies of the matings from the ecological and behavioural mechanisms of sexual selection that determine the selective advantage gained.

In Karlin and O'Donald's models, constantly expressed preferences determine both the random and assortative preferential matings. Models with partial or frequency-dependent expression of preference (in press) have recently been applied to cases of assortative mating. Table 9.1 shows the components of the matings when both dominant and recessive characters determined by two alleles at a locus are subject to both assorting and non-assorting preferences: the model combines random mating and preferential mating with and without assortment in a polygynous organism.

Let u, v and w be the frequencies of genotypes AA, Aa and aa among adults in one generation. Then in the next generation, these genotypic frequencies become

$$
\left.\begin{aligned}
u' &= p^2(\alpha+a)/(1-w) + p^2(1-\alpha-\gamma-a) \\
v' &= \alpha p(q+\tfrac{1}{2}v)/(1-w) + \gamma p + avp/(1-w) \\
&\quad + 2pq(1-\alpha-\gamma) - wp(a+c) - avp \\
w' &= \tfrac{1}{2}\alpha vq/(1-w) + \gamma q + \tfrac{1}{4}av^2/(1-w) + cw \\
&\quad + q^2(1-\alpha-\gamma) - q(cw+\tfrac{1}{2}av)
\end{aligned}\right\} \tag{9.1.1}
$$

where p and q are the gene frequencies defined as usual by

$$p = u + \tfrac{1}{2}v$$
$$q = w + \tfrac{1}{2}v$$

From the values of u' and v' in equations 9.1.1, the gene frequency in the next generation can then be shown to be given by the equation

$$p' = p[1 - \tfrac{1}{2}(\alpha+\gamma) + \tfrac{1}{2}\alpha/(1-w) + \tfrac{1}{2}w(a-c)]$$

Hence we find the following change in gene frequency from one generation to the next:

$$\Delta p = \tfrac{1}{2}p[-\gamma + w(\alpha+\gamma) + w(1-w)(a-c)]/(1-w) \tag{9.1.2}$$

If all preferential matings are assortative ($\alpha=\gamma=0$), then no stable polymorphism exists: equation 9.1.2 reduces to

$$\Delta p = \tfrac{1}{2}pw(a-c)$$

showing that AA fixes if $a>c$ and aa fixes if $a<c$.

There are two possible outcomes if either $\alpha=0$ or $\gamma=0$.

Case (i): $\alpha=0$.   In this case, either fixation of aa may occur or a stable polymorphism may be established. If $a<\gamma+c$, then aa fixes. If $a>\gamma+c$, then a polymorphism is established at a frequency of aa given by

$$w_* = \gamma/(a-c) \tag{9.1.3}$$

Case (ii): $\gamma=0$.   Now AA is fixed if $c<\alpha+a$ and a stable polymorphism is established if $c>\alpha+a$ at the frequency

$$w_* = [c-(\alpha+a)]/(c-a) \tag{9.1.4}$$

In the general case, $\alpha\gamma>0$, a globally stable equilibrium is established at which

$$\gamma - w(\alpha + \gamma) = w(1 - w)(a - c)$$

Karlin and O'Donald give the relevant solution for the frequency at equilibrium:

$$w_* = \frac{\alpha + \gamma + a - c - \sqrt{[(\alpha + \gamma + a - c)^2 - 4\gamma(a - c)]}}{2(a - c)} \qquad (9.1.5)$$

Thus in this model, the advent of random preferential mating prohibits the inevitable fixation that would occur by assortative mating alone. If $a = c > 0$, we get exactly the same polymorphism as that with random preferential mating alone ($a = c = 0$) at an equilibrium frequency

$$w_* = \gamma/(\alpha + \gamma)$$

as found in section 4.1, equation 4.1.4. However in the combined model in which $a = c > 0$, the rate of approach to this equilibrium is slower than the rate in the simple model with no assortative mating.

9.2   Separate preferences for each genotype

When females may prefer one of the three genotypes AA, Aa or aa and may express their preferences either with or without assortment, then the components of the matings occur at the frequencies shown in table 9.2. The genotypic frequencies u, v and w are then given in the next generation by the equations

$$\left.\begin{array}{l} u' = p(\alpha + \tfrac{1}{2}\beta) + au + \tfrac{1}{4}bv + p^2(1 - \theta) - p(au + \tfrac{1}{2}bv) \\[2pt] v' = p(\gamma + \tfrac{1}{2}\beta) + q(\alpha + \tfrac{1}{2}\beta) + 2pq(1 - \theta) - cwp - auq \\[2pt] w' = q(\gamma + \tfrac{1}{2}\beta) + cw + \tfrac{1}{4}bv + q^2(1 - \theta) - q(cw + \tfrac{1}{2}bv) \end{array}\right\} \qquad (9.2.1)$$

where $\theta = \alpha + \beta + \gamma$ as shown in the table.

Karlin and O'Donald (1978) gave analytical solutions to equations 9.2.1 for the symmetric case when $\alpha = \gamma$ and $a = c$. Numerical solutions can easily be obtained by computer without this restriction, which is admittedly unrealistic. However, as Karlin and O'Donald point out, the numerical results in the general case are qualitatively similar to those of the symmetric case. In the symmetric case, the equations can be represented by the pair of

Table 9.2 *Mating frequencies in the combined model with separate pre-ferences for each genotype*

	Components of matings		
Mating types	Random preferential mating	Assortative preferential mating	Random mating
$AA \times AA$	αu	au	$u^2(1-\theta-a)$
$AA \times Aa$	$\alpha v + \beta u$	0	$uv(2-2\theta-a-b)$
$AA \times aa$	$\alpha w + \gamma u$	0	$uw(2-2\theta-a-c)$
$Aa \times Aa$	βv	bv	$v^2(1-\theta-b)$
$Aa \times aa$	$\beta w + \gamma v$	0	$vw(2-2\theta-b-c)$
$aa \times aa$	γw	cw	$w^2(1-\theta-c)$

In this table α, β and γ are the parameters for random preferential mating and a, b and c the parameters for assortative mating of the genotypes AA, Aa and aa. $\theta = \alpha + \beta + \gamma$.

equations

$$u' - w' = (u-w)[1 - \alpha - \tfrac{1}{2}\beta + \tfrac{1}{2}v(a-b)]$$
$$v' = \tfrac{1}{2} - \tfrac{1}{2}(u-w)^2(1 - 2\alpha - \beta - a) - \tfrac{1}{2}a(1-v)$$

These equations have two different types of solution depending on conditions. In one case, there is a unique, globally stable equilibrium; in the other case there are two stable equilibria and one unstable equilibrium.

Case (i). If $(a-b)(1-a) + (a-2)(2\alpha+\beta) < 0$ then this condition ensures that a unique, globally stable equilibrium exists at frequencies

$$\left. \begin{aligned} u_* = w_* &= 1/[2(2-a)] \\ v_* &= (1-a)/(2-a) \end{aligned} \right\} \qquad (9.2.3)$$

These equilibrium frequencies are thus independent of the parameters of random preferential mating. They are exactly the frequencies at equilibrium with assortative mating alone: the rate of approach to the values of 9.2.3 then occurs at a geometric rate given by the multiplying factor

$$\lambda = 1 + (a-b)(1-a)/[2(2-a)]$$

When random preferential mating is superimposed on the assortative mating, the geometric rate of approach is increased to

$$\lambda = 1 + (a-b)(1-a)/[2(2-a)] - \alpha - \tfrac{1}{2}\beta$$

If none of the matings take place assortatively, then the rate of approach is simply

$$\lambda = 1 - \alpha - \tfrac{1}{2}\beta$$

or in general when $\alpha \neq \gamma$

$$\lambda = 1 - \tfrac{1}{2}\theta$$
$$= 1 - \tfrac{1}{2}(\alpha + \beta + \gamma)$$

as shown in equation 4.2.3, section 4.2. With no assortment, however, the gene frequency converges globally to the value

$$p_* = (\alpha + \tfrac{1}{2}\beta)/(\alpha + \beta + \gamma)$$

with the genotypes maintained at the Hardy–Weinberg frequencies

$$u_* = p_*^2$$
$$v_* = 2p_* q_*$$
$$w_* = q_*^2$$

The condition of case (i) is certainly satisfied when $b > a$; in the presence of random preferential mating, a more extended set of values of the parameters can satisfy the condition since the factor $(a-2)(2\alpha+\beta)$ is always negative. It is remarkable that the introduction of even the smallest degree of assortative mating can completely override the effects of random preferential mating. It is surprising too that the randomly distributed preferences have no effect on the degree of homozygosity attained at equilibrium: in the simple case without assortment, Hardy–Weinberg frequencies are maintained; we might expect, therefore, that the non-assorting preferences would reduce the degree of homozygosity; and yet the genotypic frequencies at equilibrium are exactly the same as when all preferential matings assort.

Case (ii). If the condition $(a-b)(1-a)+(a-2)(2\alpha+\beta) > 0$ holds for the values of the parameters, then the equilibrium 9.2.3 still exists, but is now unstable. The genotypic frequencies are now maintained at one of two sets of values, (u_1^*, v_1^*, w_1^*) and (u_2^*, v_2^*, w_2^*), representing the two alternative stable equilibria. If we define

$$\zeta = [1 - a + (a-2)(2\alpha+\beta)/(a-b)]/(1 - 2\alpha - \beta - a)$$

and

$$v^* = (2\alpha + \beta)/(a - b)$$

then

$$\left.\begin{array}{l} u_1^* = \tfrac{1}{2}(1 - v^* - \sqrt{\zeta}), \; v_1^* = v^*, \; w_1^* = \tfrac{1}{2}(1 - v^* + \sqrt{\zeta}) \\ u_2^* = \tfrac{1}{2}(1 - v^* + \sqrt{\zeta}), \; v_2^* = v^*, \; w_2^* = \tfrac{1}{2}(1 - v^* - \sqrt{\zeta}) \end{array}\right\} \qquad (9.2.4)$$

The convergence to one or other of these equilibria depends on the initial frequencies. If $w_0 > u_0$, (or equivalently $p_0 < \tfrac{1}{2}$), this condition then defines the domain of convergence to (u_1^*, v_1^*, w_1^*). If $u_0 > w_0$ (or $p_0 > \tfrac{1}{2}$), this defines convergence to (u_2^*, v_2^*, w_2^*). An example giving the condition for case (ii) and the stable equilibria of 9.2.4 would have numerical values $a = c = \tfrac{1}{2}$, $b = 0$, $2\alpha + \beta < 1/6$. Table 9.3, taken from Karlin and O'Donald's paper (1978), gives the genotypic frequencies at equilibrium over a range of values of $2\alpha + \beta$. When $2\alpha + \beta > 1/6$, the condition for case (i) holds giving the stable equilibrium 9.2.3.

9.3 Fitting the model to data of *Mormoniella*

Grant, Snyder and Glessner (1974) obtained experimental data on the matings of wild-type and mutant phenotypes of the parasitic wasp *Mormoniella vitripennis*. Males showed no significant preference for wild-type or mutant females, but females more frequently chose the wild-type males for mating. As we should expect, the sexual advantage was frequency-dependent. At a frequency of 0.2, wild-type males had a great advantage over the mutants; at equal frequencies of wild-types and mutants, some of this advantage was lost; mutants at a frequency of 0.2 gained some advantage over wild-types. The mutants were in fact double mutants: they carried both body colour and eye colour mutations. The mating trials were carried out with ten males and ten females. Within each trial frequencies of wild-types (+) and mutants (m) were held constant. In different trials, the males' frequencies were varied between the three ratios 2(+):8(m), 5(+):5(m) and 8(+):2(m). Mutant and wild-type females were always present at equal frequencies. Table 9.4 gives the data in the form presented by Grant, Snyder and Glessner. The assortative mating is very clear and highly significant. For example, when males at the ratio 2(+):8(m)

Table 9.3 *Equilibrium genotypic frequencies in the combined model without dominance: homozygotes have equal assorting rates (given by $a=c=\frac{1}{2}$); heterozygotes do not assort ($b=0$). Sexual selection is determined by mating preferences $\alpha=\gamma$ and β, giving a total proportion $2\alpha+\beta$ of preferential matings*

Preferential matings $2\alpha+\beta$	Genotype frequencies			Value of conditional expression $a(1-a)+(a-2)(2\alpha+\beta)$
	u^*	v^*	w^*	
0.06	0.0136	0.1200	0.8664	0.16
0.08	0.0266	0.1600	0.8134	0.13
0.10	0.0464	0.2000	0.7536	0.10
0.12	0.0765	0.2400	0.6835	0.07
0.14	0.1243	0.2800	0.5957	0.04
0.16	0.2187	0.3200	0.4613	0.01
0.18	0.3333	0.3333	0.3333	-0.02
0.20	0.3333	0.3333	0.3333	-0.05

If the value of the conditional expression $a(1-a)+(a-2)(2\alpha+\beta)<0$ then the single, globally stable equilibrium of case (1) is reached. But if the condition for case (ii) holds with $a(1-a)+(a-2)(2\alpha+\beta)>0$, then two different equilibria are stable: if $p_0<\frac{1}{2}$ then the equilibrium frequencies are given by the expressions for u_1^*, v_1^* and w_1^*; if $p_0>\frac{1}{2}$ then the equilibrium frequencies are given by u_2^*, v_2^* and w_2^*. In this table the equilibria of case (ii) are those for which $2\alpha+\beta<0.167$: only the frequencies u_1^*, v_1^* and w_1^* are shown in the table; the frequencies u_2^*, v_2^* and w_2^* can be found by interchanging the values of u^* and w^*.

are chosen for mating, we have the contingency table

females

		+	m
males	+	64	38
	m	78	126

Table 9.4 *Matings of* Mormoniella vitripennis *observed for each of three frequencies of males*

Ratio among males	Numbers of matings observed				
	Wild-type (+) ♀♀		Mutant (m) ♀♀		Total
	+♂♂	m♂♂	+♂♂	m♂♂	
2(+):8(m)	64	78	38	126	306
5(+):5(m)	101	43	63	92	299
8(+):2(m)	131	35	100	47	313

The values in this table are those given by Grant, Snyder and Glessner (1974).

giving the very significant value of $\chi_1^2 = 15.454$ (with Yates' correction) for the test of departure from frequencies of random mating. Similar tests on the matings at other frequencies also give significant values of χ^2. At the same time, it can be seen that wild-type males have a general advantage over mutants. At the frequency $2(+):8(m)$, the wild-type males have a considerable advantage: observed numbers of matings may be compared with those expected as shown in the following table.

		Males	
		+	m
Matings	Observed nos.	102	204
	Expected nos.	61.4	244.8

This gives $\chi_1^2 = 34.00$. At the frequency $5(+):5(m)$, the wild-type males retain a slight, though not significant advantage. But at the frequency $8(+):2(m)$, the comparison of observed and expected numbers of matings shows an excess of mutants who mate; for we have the table

		Males	
		+	m
Matings	Observed nos.	231	82
	Expected nos.	250.4	62.6

giving $\chi_1^2 = 7.52$ with the mutants at a significant advantage. Thus in these data we see both assortative mating and a general frequency-dependent mating advantage with an overall tendency somewhat in favour of the wild-type males.

If female phenotype is ignored, the data are then as follows:

		Males	
		+	m̂
Matings	Ratio $2(+):8(m)$	102	204
at given	Ratio $5(+):5(m)$	164	135
ratio	Ratio $8(+):2(m)$	231	82

In this form, the data can be fitted to the models with complete or frequency-dependent expression of preference (the 'C' or 'PF−' models). Similarly, the data of mutant females can be fitted to the

Table 9.5 *Fitting data of* Mormoniella *to models of random preferential mating*

Data fitted	Frequency-dependent preferences 'PF −' model		Constant preferences 'C' model	
All females	Before fitting: After fitting:	$\chi_3^2 = 44.328$ $\hat{\alpha} = 0.224$ $\hat{\gamma} = 0.103$ $\chi_1^2 = 0.489$	Before fitting: After fitting:	$\chi_3^2 = 44.328$ $\hat{\alpha} = 0.203$ $\hat{\gamma} = 0.124$ $\chi_1^2 = 0.141$
Mutant females	Before fitting: After fitting:	$\chi_3^2 = 19.626$ $\hat{a} = 0.044$ $\hat{\gamma} = 0.217$ $\chi_1^2 = 1.922$	Before fitting: After fitting:	$\chi_3^2 = 19.626$ $\hat{\alpha} = 0.072$ $\hat{\gamma} = 0.189$ $\chi_1^2 = 1.099$
Wild-type females	Before fitting: After fitting:	$\chi_3^2 = 79.265$ $\hat{\alpha} = 0.449$ $\hat{\gamma} = 0.015$ $\chi_1^2 = 6.350$	Before fitting: After fitting:	$\chi_3^2 = 79.265$ $\hat{\alpha} = 0.374$ $\hat{\gamma} = 0.086$ $\chi_1^2 = 2.927$

The model with frequency-dependent expression of preference is the model described in section 3.4. The expression of preference in this model is proportional to the frequency of those phenotypes that are not the object of the preference exercised. In the model α is the preference for wild-type males and γ the preference for mutant males.

models separately from the data of the wild-type females. Table 9.5 shows the results of these calculations. The proportion of females preferring the mutant males, γ, is greater among mutant females than among wild-type females; the proportion preferring wild-type males, α, is greater among the wild-type females. Even so, some mutant females prefer wild-type males and some wild-type females prefer mutant males. Evidently, the preferential matings are not all assortative. Both random and assortative preferential mating therefore combine to determine the overall frequencies of the matings. If the data of mutant and wild-type females are neither lumped together nor analysed separately, Karlin and O'Donald's model is then necessary to fit the data and estimate the parameters of random and assortative preferential mating.

In Grant, Snyder and Glessner's experiments, the frequencies of the males were varied while those of the females were held constant. For generality, we should consider two phenotypes, A and B, at frequencies u and v in females and x and y in males. If α and γ, a and c are the parameters of random and assortative preferential mating, then preferential matings take place at the frequencies:

		Males		Proportions of females mating at random
		A	*B*	
Females	*A*	$u(\alpha + a)$	γu	$u(1 - \alpha - \gamma - a)$
	B	αv	$v(\gamma + c)$	$v(1 - \alpha - \gamma - c)$

Let the overall frequencies of matings be symbolized as follows:

		Males	
		A	*B*
Females	*A*	$P_T(AA)$	$P_T(AB)$
	B	$P_T(BA)$	$P_T(BB)$

Then we have the following equations for the expected frequencies of the matings:

$$P_T(AA) = u(\alpha + a) + ux(1 - \alpha - \gamma - a)$$

$$P_T(AB) = \gamma u + uy(1 - \alpha - \gamma - a)$$

$$P_T(BA) = \alpha v + vx(1 - \alpha - \gamma - c)$$

$$P_T(BB) = v(\gamma + c) + vy(1 - \alpha - \gamma - c)$$

When multiplied by the total number of matings observed at each ratio of males, these expressions then give the expectations of the numbers of matings observed. If at a particular male ratio, the observed numbers of matings are $N(AA)$, $N(AB)$, $N(BA)$ and $N(BB)$, then the log. likelihood is given by the equation

$$\log L = \Sigma\{N(AA)\log[P_T(AA)] + N(AB)\log[P_T(AB)]$$

$$+ N(BA)\log[P_T(BA)] + N(BB)\log[P_T(BB)]\}$$

where the summation is taken over each of the different frequencies of males and females in the different experiments. The method of trial and error, described in section 4.5, has been used to find the maximum likelihood estimates of the parameters. The expected numbers of the matings, corresponding to the observed numbers of table 9.4, were calculated from the values of $P_T(AA)$, $P_T(AB)$, $P_T(BA)$ and $P_T(BB)$ at maximum likelihood and hence the residual χ^2 after the fitting of the model. There are four classes of matings and three degrees of freedom after the total number of matings at each male ratio has been allowed for – nine degrees of freedom before any parameters have been fitted.

Table 9.6 gives the estimates of the parameters and the analysis of χ^2. It shows that the parameters γ and a must be fitted in order to reduce the residual heterogeneity to a level of non-significance. After fitting γ and a, $\chi^2_7 = 11.324$ corresponding to $P = 0.13$. Fitting the parameters c and then α gives the values $\chi^2_6 = 10.746$ and $\chi^2_5 = 7.095$ respectively. These reductions in the values of χ^2 do not represent significant reductions in residual heterogeneity. The table also shows the equilibrium frequency ultimately attained by the mutant phenotype on the assumption that the mutant is recessive. These values were calculated from the expression for w_* derived in section 9.1. Convergence to stable equilibrium was also checked by numerical computation of the recurrence equations 9.1.1. For those values of the parameters that fit the data and leave no residual heterogeneity, the mutant phenotype would reach an equilibrium frequency of between 0.27 and 0.31 depending on which parameters had been fitted. Predictions about the equilibrium frequency would be testable in population cage experiments. In Grant, Snyder and Glessner's experiments, however, a double mutant was used, and as stated in their paper,

Although the mutant proved convenient in the observation of the various matings, we cannot assume that these phenotypic differences actually accounted for the rare-type advantage or the mating discrimination between the two types. Perhaps obvious, nonetheless, it should be pointed out that the two stocks employed are inbred lines and that the genetic background shared by the mutants may differ significantly from the genetic background shared by those designated as wild.

Separate preferences evolved within each inbred line would explain the observation of assortative mating of mutant with mutant and wild-type with wild-type. Table 9.6 shows that larger estimates are obtained for the parameters of assortative preferential mating than for those of random preferential mating. Comparison of tables 9.5 and 9.6 shows that the data of mutant females (table 9.5) give the same estimate of the preference for mutant males as the combined model with $\gamma = 0$ (table 9.6). Similarly, the data of wild-type females give the same estimate for wild-type males as the combined model with $\alpha = 0$.

We have already seen in Chapter 7 that assortative and random preferential mating are probably combined in the Arctic Skua. Since estimates of the non-assorting components of preference were derived from the distributions of the males' breeding dates, while estimates of the assorting components were derived from the

Table 9.6 *Fitting Karlin and O'Donald's combined model to the data of Mormoniella*

Parameters fitted	Maximum likelihood estimates of parameters				Log. likelihood	Residual χ^2	Dfs	Equilibrium frequency
	$\hat{\alpha}$	$\hat{\gamma}$	$\hat{a}$	$\hat{c}$				
—	—	—	—	—	−1236.76	96.643***	9	—
α	0.1198	—	—	—	−1225.18	69.532***	8	0.0
γ	—	0.0089	—	—	−1236.69	96.532***	8	1.0
a	—	—	0.3028	—	−1205.97	31.286***	8	0.0
c	—	—	—	0.1369	−1229.79	82.319***	8	1.0
α, γ	0.2027	0.1239	—	—	−1217.01	50.359***	7	0.3795
α, a	0.0000	—	0.3033	—	−1205.97	31.294***	7	0.0
α, c	0.2064	—	—	0.2768	−1204.40	25.822***	7	0.2543
γ, a	—	0.1141	0.3951	—	−1196.92	11.324	7	0.2889
γ, c	—	0.0000	—	0.1369	−1229.79	82.319***	7	1.0
a, c	—	—	0.3028	0.1369	−1198.99	16.962**	7	0.0
α, γ, a	0.0430	0.1312	0.3642	—	−1196.22	9.509	6	0.3078
α, γ, c	0.2128	0.0133	—	0.0268	−1204.32	25.429***	6	0.3134
α, a, c	0.0723	—	0.2305	0.1888	−1197.46	13.311*	6	0.0
γ, a, c	—	0.0860	0.3744	0.0509	−1196.52	10.746	6	0.2658
α, γ, a, c	0.0723	0.0860	0.3021	0.1028	−1194.99	7.095	5	0.2861

In this table the residual χ^2 is the value of χ^2 after the parameters indicated have been fitted, each parameter removing one degree of freedom (Df). Levels of significance are indicated by

***for $P < 0.01$
**for $0.01 < P < 0.02$
*for $0.02 < P < 0.05$

The equilibrium frequency is the ultimate frequency of the mutant phenotype at a stable equilibrium state given by equation 9.1.5.

overall frequencies of the matings, it would not be possible to fit the combined model to the Arctic Skua data. To do so, it would be necessary to classify the breeding dates of newly mated birds according to the phenotypes of both the parents, but there are insufficient data for this further classification of each type of male according to the type of female he was mated to.

To test the goodness of fit of the combined model to data of natural populations, large numbers of both mated and unmated individuals would have to be counted in a series of populations with different genotypic frequencies. Data of this kind will be difficult to obtain on a sufficiently large scale. However, laboratory experiments of the sort carried out by Grant, Snyder and Glessner should be applicable to a wide range of organisms. Spiess first designed such experiments (Spiess, 1968). We have already seen that his data on mating choice in *Drosophila* fit the models of partial preference very well (section 5.5). The matings were not assortative, however, and have not been fitted to the combined model.

If mating preferences have evolved by the selection of females who prefer to mate with the fittest males and whose preferences thus further increase these males' selective advantage, then some degree of assortative mating is likely to evolve as a result of the establishment of linkage disequilibrium between genes for the preferences and genes for the preferred characters (Chapter 8, sections 8.3 and 8.4). Only a low level of assortative mating will evolve (see tables 8.5 and 8.6, Chapter 8). But assortative mating will be selected further if two different phenotypes tend to occur in different ecological niches or geographical ranges. Different inbred strains will also tend to evolve preferences within strains as shown by Grant, Snyder and Glessner's data.

Any assortative mating that evolves necessarily reduces the gene flow between strains or races, thus increasing the advantage of assortative preferential mating over random preferential mating: the preference is more closely associated with the preferred genotype and is therefore selected more rapidly. We may expect that randomly distributed preferences will evolve first; but, as the expression of preference becomes partly assortative, so the assortative preference will acquire its own advantage: Fisher's 'runaway process' of sexual selection may thus become to some extent a runaway process of assortative mating with selection for mechanisms to increase the expression of preference among preferred

Table 9.7 *'Encounter models' of partial sexual selection and assortative mating: females mate at random after a certain number of encounters with courting males unless they have met a male they prefer*

Geno- types	Ran- dom mating	Proportions of preferential matings determined by numbers of encounters before females mate at random		
		Effect of random preferential matings		
		$m=1$	$m=2$	$m \to \infty$
AA	p^2	$p^2 w(\alpha - \gamma)$	$p^2 w[\alpha - 2\gamma + w(\alpha + \gamma)]$	$\dfrac{\alpha p^2 w - \gamma p^2 (1-w)}{1-w}$
Aa	$2pq$	$pw(q-p)(\alpha - \gamma)$	$pw(q-p)[\alpha - 2\gamma + w(\alpha + \gamma)]$	$\dfrac{p(q-p)[\alpha w - \gamma(1-w)]}{1-w}$
aa	q^2	$-pqw(\alpha - \gamma)$	$-pqw[\alpha - 2\gamma + w(\alpha + \gamma)]$	$\dfrac{-pq[\alpha w - \gamma(1-w)]}{1-w}$
Total	1	0	0	0

In this table the values of m represent the numbers of encounters that a female will require before she mates at random if her preference is not expressed assortatively; n is the corresponding number of encounters if a female's preference is expressed assortatively in association with the genotypes that determine the male phenotype she prefers.

genotypes. This process might therefore lead to further divergence and increasing isolation, yet without the existence of any initial geographical or even ecological barriers to matings.

9.4 Partial preferences for dominant and recessive characters

Karlin and O'Donald's model of sexual selection and assortative mating can easily be extended to allow for partial expression of preference. As in section 9.1, females are assumed to exercise their mating preferences either assortatively, in association with the preferred genotypes the females themselves carry, or non-assortatively, without regard to these genotypes: α and γ are the random or non-assortative preferences for the dominant character A (genotypically AA or Aa) and the recessive aa; a and c are the corresponding assortative preferences. In the 'encounter models' of partial preference, females only express their preferences if they meet a male they prefer within a particular number of successive encounters with courting males; otherwise, they mate with the next male they meet regardless of his phenotype. In the general model, a female with a non-assortative preference mates at random if she has

Table 9.7 *continued*

Effect of assortative preferential matings		
$n=1$	$n=2$	$n \to \infty$
ap^2w	$ap^2w(1+w)$	$\dfrac{ap^2w}{1-w}$
$-apuw-cpw^2$	$-apuw(1+w)-cpw^2(2-w)$	$\dfrac{-apuw-cpw(1-w)}{1-w}$
$-\frac{1}{2}apvw+cpw^2$	$-\frac{1}{2}apvw(1+w)+cpw^2(2-w)$	$\dfrac{-\frac{1}{2}apvw+cpw(1-w)}{1-w}$
0	0	0

encountered m courting males, none of whom has the phenotype she prefers. If her preference is assortative and exercised only if she carries a genotype that determines the character she prefers, she mates at random after n encounters with males who possess the wrong phenotype. Table 9.7 shows the genotypic frequencies after selection when the parameters m and n take values 1 or 2, or are increased without limit to give rise to complete expression of the preferences. To find the recurrence equations for any particular case, the effects of random and assortative preferential mating are added to the simple random mating frequencies p^2, $2pq$ and q^2 of the genotypes AA, Aa and aa. For example, when $m=n=1$, we have the equations

$$u' = p^2 + p^2w(\alpha - \gamma) + ap^2w$$
$$v' = 2pq + pw(q-p)(\alpha - \gamma) - apuw - cpw^2$$
$$w' = q^2 - pqw(\alpha - \gamma) - \tfrac{1}{2}apvw + cpw^2$$

Since $u+v+w=1$, only two variables are necessary to describe the models fully. In this case, as in Karlin and O'Donald's model with complete expression of preference, the equation for change in gene frequency gives the equilibrium frequency of the aa phenotype:

$$p' = p + \tfrac{1}{2}pw[\alpha - \gamma + a(1-w) - cw]$$

At equilibrium $p' = p$ and hence equilibria exist at the frequencies

$$w_* = 0, \ 1, \ (\alpha - \gamma + a)/(a + c)$$

The recurrence equation can therefore be written in the form

$$p' = p + \tfrac{1}{2}pw(w_* - w)(a + c)$$

Thus if $w > w_*$, aa fixes since $\Delta p < 0$ and $w_{i+1} > w_i$. If $w < w_*$, AA fixes since $\Delta p > 0$ and $w_{i+1} < w_i$. The polymorphic equilibrium at the frequency w_* is therefore unstable.

Table 9.8 shows the recurrence equations for gene and phenotype frequencies for the nine sets of values:

$$m = 1; \ n = 1, \ n = 2, \ n \to \infty$$

$$m = 2; \ n = 1, \ n = 2, \ n \to \infty$$

$$m \to \infty; \ n = 1, \ n = 2, \ n \to \infty$$

The existence of polymorphic equilibria can be found by putting $p' = p$ and solving the polynomial equation for w. Generally we may expect that in all cases when $m = 1$, the equilibria will be unstable. This follows because with assortative mating alone fixation always takes place: if the frequencies of the genotypes are represented by (u, v, w), the two fixation states $(1, 0, 0)$ and $(0, 0, 1)$ are both locally stable for the case α, $\gamma = 0$, $ac > 0$ (Raper, Karlin and O'Donald, 1979). We have already seen that when $m = n = 1$ in the model with both assorting and non-assorting parameters, both states $(1, 0, 0)$ and $(0, 0, 1)$ are locally stable while the polymorphic state is unstable: the non-assortative preferences give rise to no frequency-dependence in the mating advantage gained; hence there is nothing to prevent the fixation that takes place by assortative mating alone. When $m = 1$ and $n = 2$, there is an equilibrium given by

$$\alpha - \gamma + a - 2cw_* - w_*^2(a - c) = 0$$

The recurrence equation for the gene frequency can then be written as

$$p' = p + \tfrac{1}{2}pw(w_* - w)[2c + 2w_*(a - c)]$$
$$- \tfrac{1}{2}pw(w_* - w)^2(a - c)$$

Near the equilibrium value w_*, the last term is negligible and we have

$$p' = p + pw(w_* - w)[c(1 - w_*) + aw_*]$$

Table 9.8 *Recurrence equations for the 'encounter models' of partial sexual selection and assortative mating*

(i) *One encounter is sufficient to produce random mating of females whose preferences are expressed non-assortatively* ($m = 1$)

Encounters when preferences
are expressed assortatively

$n = 1$
$$\begin{cases} p' = p + \tfrac{1}{2}pw[(\alpha - \gamma) + a(1 - w) - cw] \\ w' = q^2 - pqw(\alpha - \gamma) - \tfrac{1}{2}apvw + cpw^2 \end{cases}$$

$n = 2$
$$\begin{cases} p' = p + \tfrac{1}{2}pw[(\alpha - \gamma) + a(1 - w^2) - cw(2 - w)] \\ w' = q^2 - pqw(\alpha - \gamma) - \tfrac{1}{2}apvw(1 + w) + cpw^2(2 - w) \end{cases}$$

$n \to \infty$
$$\begin{cases} p' = p + \tfrac{1}{2}pw(\alpha - \gamma) + \tfrac{1}{2}pw(a - c) \\ w' = q^2 - pqw(\alpha - \gamma) - pw[\tfrac{1}{2}av - c(1 - w)]/(1 - w) \end{cases}$$

(ii) *Two encounters are required to produce random mating of females whose preferences are expressed non-assortatively* ($m = 2$)

Encounters when preferences
are expressed assortatively

$n = 1$
$$\begin{cases} p' = p + \tfrac{1}{2}pw[\alpha - 2\gamma + w(\alpha + \gamma) + a(1 - w) - cw] \\ w' = q^2 - pqw[\alpha - 2\gamma + w(\alpha + \gamma)] - \tfrac{1}{2}apvw + cpw^2 \end{cases}$$

$n = 2$
$$\begin{cases} p' = p + \tfrac{1}{2}pw[\alpha - 2\gamma + w(\alpha + \gamma) + a(1 - w^2) - cw(2 - w)] \\ w' = q^2 - pqw[\alpha - 2\gamma + w(\alpha + \gamma)] - \tfrac{1}{2}apvw(1 + w) + cpw^2(2 - w) \end{cases}$$

$n \to \infty$
$$\begin{cases} p' = p + \tfrac{1}{2}pw[\alpha - 2\gamma + w(\alpha + \gamma) + a - c] \\ w' = q^2 - pqw[\alpha - 2\gamma + w(\alpha + \gamma)] - [\tfrac{1}{2}apvw - cpw(1 - w)]/(1 - w) \end{cases}$$

(iii) *Females always mate preferentially when their preferences are expressed non-assortatively* ($m \to \infty$)

Encounters when preferences
are expressed assortatively

$n = 1$
$$\begin{cases} p' = p + \tfrac{1}{2}p[-\gamma + w(\alpha + \gamma) + aw(1 - w)^2 - cw^2(1 - w)]/(1 - w) \\ w' = q^2 - pq[\alpha w - \gamma(1 - w)]/(1 - w) - \tfrac{1}{2}apvw + cpw^2 \end{cases}$$

$n = 2$
$$\begin{cases} p' = p + \tfrac{1}{2}p[-\gamma + w(\alpha + \gamma) + aw(1 - w^2)(1 - w) - cw^2(2 - w)(1 - w)]/(1 - w) \\ w' = q^2 - pq[\alpha w - \gamma(1 - w)]/(1 - w) - \tfrac{1}{2}apvw(1 + w) + cpw^2(2 - w) \end{cases}$$

$n \to \infty$
$$\begin{cases} p' = p + \tfrac{1}{2}p[-\gamma + w(\alpha + \gamma) + w(1 - w)(a - c)]/(1 - w) \\ w' = q^2 - \{pq[\alpha w - \gamma(1 - w)] + \tfrac{1}{2}apvw - cpw(1 - w)\}/(1 - w) \end{cases}$$

Since the expression $c(1 - w_*) + aw_*$ is always positive, the equilibrium point w_* is unstable. When $m = 1$ and $n \to \infty$ there is no polymorphic equilibrium:

$$p' = p + \tfrac{1}{2}pw(\alpha + a - \gamma - c)$$

showing that *AA* fixes when $\alpha + a > \gamma + c$ and otherwise *aa* fixes. Thus in general when $m = 1$, no polymorphism can be maintained by a combination of assorting and non-assorting preferential matings.

In models in which $m \geq 2$, we may expect that stable polymorphisms will exist: the mating advantage will have a component of negative frequency-dependence that may be sufficient to prevent fixation from taking place.

Model with m = 2, n = 1

The recurrence equation in table 9.8 shows that a polymorphic equilibrium can exist at a frequency

$$w_* = (2\gamma - \alpha - a)/(\alpha + \gamma - a - c)$$

Thus we have in terms of w_*, the recurrence equation as follows:

$$p' = p + \tfrac{1}{2}pw(w - w_*)(\alpha + \gamma - a - c)$$

Then, on the condition that $\alpha + \gamma > a + c$, if $w > w_*$, $\Delta p > 0$ and $w_{i+1} < w_i$; while if $w < w_*$, $\Delta p < 0$ and $w_{i+1} > w_i$. Therefore w_* is a stable point; it represents a polymorphism if $2\gamma - a > \alpha > \tfrac{1}{2}(\gamma + c)$.

Model with m = 2, n = 2

As shown by the recurrence equation in table 9.8, there is an equilibrium at a frequency given by the equation

$$\alpha - 2\gamma + a + w_*(\alpha + \gamma) - aw_*^2 - 2cw_* + cw_*^2 = 0$$

Then for the recurrence equation in terms of w_*, we have

$$p' = p + \tfrac{1}{2}pw(w - w_*)[\alpha + \gamma - 2aw_* - 2c(1 - w_*)]$$
$$- \tfrac{1}{2}pw(w - w_*)^2(a - 2c)$$

and so near equilibrium

$$p' = p + \tfrac{1}{2}pw(w - w_*)[\alpha + \gamma - 2aw_* - 2c(1 - w_*)]$$

The equilibrium is stable if $\alpha + \gamma > 2[aw_* + c(1 - w_*)]$. If $a = c$, this is the same condition as that for stability when $m = 2$ and $n = 1$.

Model when m = 2, n → ∞

The recurrence equation gives the equation for equilibrium

$$w_* = (2\gamma - \alpha - a + c)/(\alpha + \gamma)$$

so that

$$p' = p + \tfrac{1}{2}pw(w - w_*)(\alpha + \gamma)$$

The polymorphic equilibria are clearly stable in this case; they exist provided that $2\gamma - (a - c) > \alpha > \tfrac{1}{2}\gamma - \tfrac{1}{2}(a - c)$.

At larger values of m equilibria are generally stable. In the limiting case $m, n → ∞$, the model becomes identical to that with complete preferences already analysed in section 9.1.

10

Conclusions

In the previous chapters I have described and analysed some genetic models of sexual selection. These models are hypotheses about selective processes operating in natural or laboratory populations. Genetic analysis of the models reveals their evolutionary effects. To be accepted as satisfactory explanations of evolution, they must both describe observations of mating behaviour and predict the consequent evolution. We have no reason to believe that any scientific model is absolutely true and can only hope that some models may be nearer to the truth than others: within defined limits of error, data may be compatible only with certain models; other models may thus be refuted.

Observations concerned with evolution are generally of two different kinds. We may observe the outcomes of evolutionary processes in many different species. This may produce generalizations, for example about associations of sexual differences with particular mating systems, that may qualitatively conform with the outcomes in some models but not others. Thus, as we have seen, sexual dimorphism in primates is usually found in association with polygyny, not with monogamy (Clutton–Brock, Harvey and Rudder, 1977). Since the conditions at the start of the evolution cannot be known, however, the models cannot be fitted to data of particular species. We may also observe selection acting in a natural or laboratory population, estimate the parameters of a model of this selection, and then predict the changes that should take place in subsequent generations and the equilibrium that should finally be attained. Often it may be possible to estimate some of the parameters of natural or sexual selection, but not possible to observe a natural population over a period long enough to compare actual with predicted changes. In population cages, each generation can be examined separately; parameters can be estimated using data of individuals from the same population; and the predictions of the

models compared with independent samples from successive generations. The models are refuted if they do not predict the actual
changes in the population cage, but this does not imply that the
selection in the model is not operating in the population: it may
only imply that other selective processes are operating in addition
to those that were postulated and measured. This problem is more
serious when predictions are applied to natural populations: selection may well be different in different environments; deviations
from a predicted set of equilibrium frequencies, for example, are
only to be expected when several natural populations are examined. Qualitative agreement is then the most that can be achieved;
models cannot be decisively refuted on the grounds of significant
quantitative deviation from prediction. Fortunately, observations
of mate selection may be sufficient in themselves for a decisive
refutation: a model is refuted if significant heterogeneity is left after
it has been fitted to data. We have seen in the sections on fitting the
models, that significant heterogeneity remains after fitting the
different models to Ehrman's data on *Drosophila*. The deviations
from the models are not consistent, however; no deterministic
model of frequency-dependent selection would explain the observed
differences in mating success at different frequencies. Sampling
variation in the level of female preference is a plausible additional
hypothesis that would explain the residual heterogeneity.

Observations of frequencies of matings, such as those of Ehrman
and Spiess, refute particular models according to the relative
advantage gained by the males at different frequencies. All models
of sexual selection by female choice must give rise to frequency-
dependent matings of the males. This is obvious if the females
always express their preferences. The proportion of preferential
matings then remains constant; a preferred male is chosen more
often when he is rare than when he is common. In the 'encounter
models' with partial expression of preference, females who have not
met a male of their choice after a certain number of encounters
with courting males eventually give up the search and mate with
the next male they meet. I have assumed that these encounters
provide stimuli that raise a female's level of stimulation from a
threshold at which she will mate only with a male she prefers up to
a threshold at which she will mate with any male. As I have shown
in Chapter 5, the advantage of being a rare male declines as the
difference in the two thresholds is reduced and females accept fewer
disappointments before mating at random. The advantage of being

rare finally disappears when females allow themselves only one disappointment before taking the next male, whoever he may be. Experiments designed to test the mating advantage gained by males at different frequencies have usually revealed a significant frequency-dependent effect. In experiments with wild-type and mutant males (Petit, 1954; Grant, Snyder and Glessner, 1974), the wild-types may have an overall advantage, but their advantage is progressively reduced as their frequency increases. Provided at least some females prefer different male phenotypes to others, the conditions for the establishment of a balanced polymorphism will be satisfied. Sexual selection alone is then sufficient to maintain the polymorphism, though natural selection may alter the point of equilibrium. Alternatively, polymorphisms can be maintained by a balance between sexual selection acting in favour of a sexually desirable character and natural selection acting against it: a polymorphism will be established on the condition that the sexual selection provides an initial advantage that is lost as the preferred males increase in frequency.

Polymorphisms are not an inevitable outcome of sexual selection, however. If females express complete preferences for more than one phenotype or genotype, then polymorphisms always exist and are globally stable. But multiple preferences do not necessarily evolve: Fisher's model of the evolution of preferences contains no premise that entails the evolution of preferences for different phenotypes; no general prediction can be made about the levels of heterozygosity that might be maintained by sexual selection. Moreover, in the more realistic models with partial expression of preference, polymorphisms are generally unstable unless there is some preference for heterozygotes; in models of the evolution of separate preferences for homozygotes and heterozygotes, the heterozygote preference is almost always eliminated. One homozygote or the other thus eventually becomes fixed in the population. Similarly, although natural selection may often balance sexual selection at a point of stable polymorphic equilibrium, this is not an inevitable outcome: natural selection may completely outweigh the effects of sexual selection; Fisher's runaway process of sexual selection may therefore come to a halt, not at a polymorphism, but at the point when there is no advantage in the selection of new alleles and hence no further development in the sexually advantageous character. Sexual selection can produce many different results – fixation of different homozygotes or polymorphisms with one or more stable equilibria – results which apply only to particular genetic systems. There is no 'General Theory of Evolution by Natural and Sexual Selection' with universal principles which can always be used to predict the outcome of selection.

The models may share certain qualitative features, however, such as female preferences acting in favour of different genotypes. The generality of these features can only be established by study of many different polymorphisms. The guppy is polymorphic for several different secondary sexual colours (Farr, 1977). Rare or novel phenotypes of male guppies show a clear advantage. Female preference would give rise to this frequency-dependent mating advantage and thus maintain the polymorphisms. The Arctic Skua is polymorphic for melanic phenotypes. Melanic males have an increased chance of finding a mate but a lower chance of surviving to breed: sexual and natural selection maintains the polymorphism. Pheasants have many different geographical races and are highly variable in plumage for display. These polymorphisms and many others can be explained by female preferences for different phenotypes. Selander (1972) in his review of sexual selection and dimorphism in birds concluded that 'the theory of sexual selection is essentially correct as stated by Darwin'. Analysis of particular polymorphisms supports this inductive conclusion. Much further research is still required to establish how widely the theory may be applied as an explanation of genetic polymorphism.

References

Allen, J. A. and Clarke, B. (1968). Evidence for apostatic selection by wild passerines. *Nature* **220**, 501–2.

Averhoff, W. W. and Richardson, R. H. (1976). Multiple pheromone system controlling mating in *Drosophila melanogaster*. *Proc. Nat. Acad. Sci. USA* **73**, 591–3.

Bastock, M. and Manning, A. (1955). The courtship of *Drosophila melanogaster*. *Behaviour* **8**, 85–111.

Bateman, A. J. (1948). Intra-sexual selection in *Drosophila*. *Heredity* **2**, 349–68.

Berry, R. J. and Davis, P. E. (1970). Polymorphism and behaviour in the Arctic Skua (*Stercorarius parasiticus* (L.)). *Proc. Roy. Soc. Lond.* B **175**, 255–67.

Burnet, B. and Connolly, K. (1973). The visual component in the courtship of *Drosophila melanogaster*. *Experientia* **29**, 488–9.

Campbell, B. (ed.) (1972). *Sexual Selection and the Descent of Man 1871–1971*. London: Heinemann.

Charlesworth, D. and Charlesworth, B. (1975). Sexual selection and polymorphism. *Amer. Natur.* **109**, 465–70.

Clutton-Brock, T. H., Harvey, P. H. and Rudder, B. (1977). Sexual dimorphism, socionomic sex ratio and body weight in primates. *Nature* **269**, 797–800.

Crook, J. H. (1962). The adaptive significance of pair formation types in weaver birds. *Symp. Zool. Soc. Lond.* **8**, 57–70.

Crook, J. H. (1965). The adaptive significance of avian social organizations. *Symp. Zool. Soc. Lond.* **14**, 181–218.

Darwin, C. R. (1859). *On the Origin of Species by Means of Natural Selection, or the Preservation of Favoured Races in the Struggle for Life*. London: John Murray.

Darwin, C. R. (1871). *The Descent of Man, and Selection in Relation to Sex*. London: John Murray.

Davis, J. W. F. and O'Donald, P. (1976a). Estimation of assortative mating preferences in the Arctic Skua. *Heredity* **36**, 235–44.

Davis, J. W. F. and O'Donald, P. (1976b). Territory size, breeding time and mating preference in the Arctic Skua. *Nature* **260**, 774–5.

Davis, J. W. F. and O'Donald, P. (1976c). Sexual selection for a handicap: a critical analysis of Zahavi's model. *J. Theor. Biol.* **57**, 345–54.

240   References

Dewar, D. and Finn, F. (1909). *The Making of Species*. London: The Bodley Head.

Eanes, W. F., Gaffney, P. M., Koehn, R. K. and Simon, C. M. (1977). A study of sexual selection in natural populations of the milkweed beetle, *Tetraopes tetraophthalmus*. In *Lecture Notes in Biomathematics* 19 *Measuring Selection in Natural Populations*, ed. F. B. Christiansen and T. M. Fenchel, pp. 49–64. Berlin: Springer-Verlag.

Ehrman, L. (1967). Further studies on genotype frequency and mating success in *Drosophila*. *Amer. Natur.* **101**, 415–24.

Ehrman, L. (1968). Frequency-dependence of mating success in *Drosophila pseudoobscura*. *Genet. Res. Camb.* **11**, 135–40.

Ehrman, L. (1969). The sensory basis of mate selection in *Drosophila*. *Evolution* **23**, 59–64.

Ehrman, L. (1970). The mating advantage of rare males in *Drosophila*. *Proc. Nat. Acad. Sci. USA* **65**, 345–8.

Ehrman, L. (1972). Genetics and sexual selection. In *Sexual Selection and the Descent of Man 1871–1971*, ed. B. Campbell, pp. 105–35. London: Heinemann.

Ehrman, L. and Spiess, E. B. (1969). Rare type mating advantage in *Drosophila*. *Amer. Natur.* **103**, 675–80.

Emlen, J. T. (1957). Display and mate selection in the Whydahs and Bishop Birds. *Ostrich* **28**, 202–13.

Farr, J. A. (1977). Male rarity or novelty, female choice behaviour, and sexual selection in the guppy, *Poecilia reticulata* Peters (*Pisces: Poeciliidae*). *Evolution* **31**, 162–8.

Fisher, R. A. (1930). *The Genetical Theory of Natural Selection*. Oxford: Clarendon Press.

Fisher, R. A. (1932). The evolutionary modification of genetic phenomena. *Proc. Sixth Int. Congr. Genet.* **1**, 165–72.

Ghai, G. L. (1974). Analysis of some non-random mating models. *Theor. Popul. Biol.* **6**, 76–91.

Grant, B., Snyder, G. A. and Glessner, S. F. (1974). Frequency-dependent mate selection in *Mormoniella vitripennis*. *Evolution* **28**, 259–64.

Hanson, H. M. (1959). Effects of discrimination training on stimulus generalization. *J. Exp. Psychol.* **58**, 321–34.

Huxley, J. S. (1938a). The present standing of the theory of sexual selection. In *Evolution: Essays on Aspects of Evolutionary Biology*, ed. G. R. de Beer, pp. 11–42. Oxford: Clarendon Press.

Huxley, J. S. (1938b). Darwin's theory of sexual selection and the data subsumed by it, in the light of recent research. *Amer. Natur.* **72**, 416–33.

Johns, J. E. (1964). Testosterone-induced nuptial feathers in phalaropes. *Condor* **66**, 449–55.

Karlin, S. (1969). *Equilibrium Behaviour of Population Genetic Models with Non-random Mating*. New York: Gordon and Breach.

Karlin, S. and O'Donald, P. (1978). Some population genetic models combining sexual selection with assortative mating. *Heredity* **41**, 165–74.

Karlin, S. and Scudo, F. M. (1969). Assortative mating based on phenotype. II. Two autosomal alleles without dominance. *Genetics* **63**, 499–510.

Kojima, K. (1971). Is there a constant fitness value for a given genotype? No! *Evolution* **25**, 281–5.

Lack, D. (1954). *The Natural Regulation of Animal Numbers.* Oxford: Clarendon Press.

Lack, D. (1966). *Population Studies of Birds.* Oxford: Clarendon Press.

Lack, D. (1968). *Ecological Adaptations for Breeding in Birds.* London: Chapman and Hall.

Lewontin, R. C. (1977). Caricature of Darwinism. *Nature* **266**, 283–4.

Maynard Smith, J. (1956). Fertility, mating behaviour, and sexual selection in *Drosophila subobscura. J. Genet.* **54**, 261–79.

Maynard Smith, J. (1958). Sexual selection. In *A Century of Darwin,* ed. S. A. Barnett, pp. 231–44. London: Heinemann.

Maynard Smith, J. (1976). Sexual selection and the handicap principle. *J. Theor. Biol.* **57**, 239–42.

Mayr, E. (1972). Sexual selection and natural selection. In *Sexual Selection and the Descent of Man* 1871–1971, ed. B. Campbell, pp. 87–104. London: Heinemann.

Merrell, D. J. (1949). Selective mating in *Drosophila melanogaster. Genetics* **34**, 370–89.

Merrell, D. J. (1950). Measurement of sexual isolation and selective mating. *Evolution* **4**, 326–31.

Miller, D. D., Goldstein, R. B. and Patty, R. A. (1975). Semispecies of *Drosophila athabasca* distinguishable by male courtship sounds. *Evolution* **29**, 531–44.

Moodie, G. E. E. (1972). Predation, natural selection and adaptation in an unusual Threespine Stickleback. *Heredity* **28**, 155–67.

Murton, R. K. and Westwood, N. J. (1977). *Avian Breeding Cycles.* Oxford: Clarendon Press.

Murton, R. K., Westwood, N. J. and Thearle, R. J. P. (1973). Polymorphism and the evolution of a continuous breeding season in the pigeon, *Columbia livia. J. Reprod. Fert. Suppl.* **19**, 563–77.

O'Donald, P. (1960). Assortative mating in a population in which two alleles are segregating. *Heredity* **15**, 389–96.

O'Donald, P. (1962). The theory of sexual selection. *Heredity* **17**, 541–52.

O'Donald, P. (1963). Sexual selection for dominant and recessive genes. *Heredity* **18**, 451–7.

O'Donald, P. (1967). A general model of sexual and natural selection. *Heredity* **22**, 499–518.

O'Donald, P. (1972a). Natural selection of reproductive rates and breeding times and its effect on sexual selection. *Amer. Natur.* **106**, 368–79.

O'Donald, P. (1972b). Sexual selection by variation in fitness at breeding time. *Nature* **237**, 349–51.

O'Donald, P. (1973a). Frequency-dependent sexual selection as a result of variation in fitness at breeding time. *Heredity* **30**, 351–68.

O'Donald, P. (1973b). Models of sexual and natural selection in polygynous species. *Heredity* **31**, 145–56.

O'Donald, P. (1974). Polymorphisms maintained by sexual selection in monogamous species of birds. *Heredity* **32**, 1–10.

O'Donald, P. (1976). Mating preferences and their genetic effects in models of sexual selection for colour phases of the Arctic Skua. In *Population Genetics and Ecology*, ed. S. Karlin and E. Nevo, pp. 411–30. New York: Academic Press.

O'Donald, P. (1977a). The mating advantage of rare males in models of sexual selection. *Nature* **267**, 151–4.

O'Donald, P. (1977b). Sexual selection and the evolution of territoriality in birds. In *Lecture Notes in Biomathematics 19 Measuring Selection in Natural Populations*, ed. F. B. Christiansen and T. M. Fenchel, pp. 113–29. Berlin: Springer-Verlag.

O'Donald, P. (1977c). Theoretical aspects of sexual selection. *Theor. Popul. Biol.* **12**, 298–334.

O'Donald, P. (1978a). Theoretical aspects of sexual selection: a generalized model of mating behaviour. *Theor. Popul. Biol.* **13**, 226–43.

O'Donald, P. (1978b). Rare male mating advantage. *Nature* **272**, 189.

O'Donald, P. (1978c). A general model of mating behaviour with natural selection and female preference. *Heredity* **40**, 427–38.

O'Donald, P. and Davis, J. W. F. (1976). A demographic analysis of the components of selection in a population of Arctic Skuas. *Heredity* **36**, 343–50.

O'Donald, P. and Davis, J. W. F. (1977). Mating preferences and sexual selection in the Arctic Skua. III. Estimation of parameters and tests of heterogeneity. *Heredity* **39**, 121–32.

O'Donald, P. and Pilecki, C. (1970). Polymorphic mimicry and natural selection. *Evolution* **24**, 395–401.

O'Donald, P., Wedd, N. S. and Davis, J. W. F. (1974). Mating preferences and sexual selection in the Arctic Skua. *Heredity* **33**, 1–16.

Parker, G. A. (1970). Sperm competition and its evolutionary effect on copula duration in the fly, *Scatophaga stercoraria*. *J. Insect Physiol.* **16**, 1301–28.

Parker, G. A. (1974a). The reproductive behaviour and the nature of sexual selection in *Scatophaga stercoraria* L. (Diptera: Scatophagidae). VIII. The behaviour of searching males. *J. Ent.* (A) **48**, 199–211.

Parker, G. A. (1974b). The reproductive behaviour and the nature of sexual selection in *Scatophaga stercoraria* L. (Diptera: Scatophagidae). IX. Spatial distribution of fertilization rates and evolution of male search strategy within the reproductive area. *Evolution* **28**, 93–108.

Petit, C. (1954). L'isolement sexuel chez *Drosophila melanogaster*. Etude du mutant white et de son allelomorphe sauvage. *Bull. Biol. France et Belgique* **88**, 435–43.

Petit, C. and Ehrman, L. (1969). Sexual selection in *Drosophila*. In *Evolutionary Biology*, ed. T. Dobzansky, M. K. Hecht and W. C. Steere, vol. 3, pp. 177–223. Amsterdam: North-Holland Publishing Company.

Raper, J. K., Karlin, S. and O'Donald, P. (1979). An assortative mating encounter model. (In press).

Reed, S. C. and Reed, E. W. (1950). Natural selection in laboratory populations of *Drosophila*. II. Competition between a white-eye gene and its wild-type allele. *Evolution* **4**, 34–42.

Scudo, F. M. and Karlin, S. (1969). Assortative mating based on phenotype. I. Two alleles with dominance. *Genetics* **63**, 479–98.

Selander, R. K. (1972). Sexual selection and dimorphism in birds. In *Sexual Selection and the Descent of Man 1871–1971*, ed. B. Campbell, pp. 180–230. London: Heinemann.

Semler, D. E. (1971). Some aspects of adaptation in a polymorphism for breeding colours in the Threespine Stickleback (*Gasterosteus aculeatus*). *J. Zool. Lond.* 165, 291–302.

Spiess, E. B. (1968). Low frequency advantage in mating of *Drosophila pseudoobscura* karyotypes. *Amer. Natur.* 102, 363–79.

Spiess, E. B. and Ehrman, L. (1978). Rare male mating advantage. *Nature* 272, 188–9.

Spiess, L. D. and Spiess, E. B. (1969). Minority advantage in interpopulational matings of *Drosophila persimilis*. *Amer. Natur.* 103, 155–72.

Staddon, J. E. R. (1975). A note on the evolutionary significance of "supernormal stimuli". *Amer. Natur.* 109, 541–5.

Sturtevant, A. H. (1915). Experiments on sex recognition and the problem of sexual selection. *J. Anim. Behav.* 5, 351–66.

Tan, C. C. (1946). Genetics of sexual isolation between *Drosophila pseudoobscura* and *Drosophila persimilis*. *Genetics* 31, 558–73.

Tebb, G. and Thoday, J. M. (1956). Reversal of mating preference by crossing strains of *Drosophila melanogaster*. *Nature* 177, 707.

Thoday, J. M. (1960). Effects of disruptive selection. III. Coupling and repulsion. *Heredity* 14, 35–49.

Thoday, J. M. and Gibson, J. B. (1962). Isolation by disruptive selection. *Nature* 193, 1164–6.

Tinbergen, N. (1948). Social releasers and the experimental method required for their study. *Wilson Bull.* 60, 6–51.

Trivers, R. L. (1972). Parental investment and sexual selection. In *Sexual Selection and the Descent of Man 1871–1971*, ed. B. Campbell, pp. 136–79, London: Heinemann.

Watson, A. (1970). Territorial and reproductive behaviour of Red Grouse. *J. Reprod. Fert. Suppl.* 11, 3–14.

Watson, A. and Moss, R. (1971). Spacing as affected by territorial behaviour, habitat and nutrition in Red Grouse (Lagopus s. scoticus). In *Behaviour and Environment: the Use of Space by Animals and Men*, ed. A. H. Esser, pp. 92–111. New York: Plenum Press.

Wilson, E. O. (1975). *Sociobiology*. Massachusetts: Belknap Press.

Witschi, E. (1961). Sex and secondary sexual characters. In *Biology and Comparative Physiology of Birds*, ed. A. J. Marshall, vol. 2, pp. 115–68. New York: Academic Press.

Zahavi, A. (1975). Mate selection – a selection for a handicap. *J. Theor. Biol.* 53, 205–14.

Index

Index